—ADVANCED LEVEL—
STATISTICS

D.A. Bryars, Ph.D.

Deputy Headmaster
Wickersley Comprehensive School
Wickersley

UNWIN HYMAN

Published by Unwin Hyman Limited
15/17 Broadwick Street
London W1V 1FP

ISBN: 0 7135 2803 6

Printed in Great Britain by
Mackays of Chatham PLC, Chatham, Kent

—ADVANCED LEVEL—

STATISTICS

Contents

Preface

The number of students following a course in Statistics has increased dramatically in recent years. This is due, in part, to the changing emphasis in the syllabuses of related A levels such as Geography and Biology, but mainly to the recognition of the importance of the subject.

This text has been written in the belief that it will be of value to all those studying A level Mathematics with a substantial Statistics component, including the revised syllabuses recently introduced by some of the Examining Boards. I trust that it will also prove to be of interest to those who are working for one of the early professional examinations.

The needs of students have been constantly borne in mind and I have endeavoured to produce a clear, concise text which develops the subject logically and at the appropriate pace. The text is punctuated by a liberal supply of worked examples and exercises. Some of the questions are designed to act as an introduction to the following section, others develop the theory further; but all are offered in the hope that they will help the student to understand and enjoy the subject.

It is, I believe, intended by the Examining Boards that Statistics, and where appropriate Mechanics, should be seen as Applied Mathematics. That is, as subjects designed to flesh out and provide motivation for the Pure Mathematics. It is, perhaps unfortunate that so many schools and colleges are obliged to split the Pure and Statistics between two teachers, so the two courses are run in parallel. To some extent this has dictated the order of the chapters. In particular, the Calculus is often not fully developed until quite late in the Pure course and so I have ensured that questions do not demand more skills than can be expected, in the early part of the text. Further, the early chapters assume little mathematical knowledge.

The first thirteen chapters can be thought of as the common core of the subject and subsequent chapters are appropriate for particular syllabuses. The exception is Chapter 15, which is more practically orientated and recognises that Statistics is also, and increasingly, seen as an applied science. It is hoped that the work in that chapter will be introduced and developed throughout the course, and the teacher should feel encouraged to introduce his own projects. However the text has been designed to be self-sufficient and should readily convert to a teaching syllabus. It is also hoped that the revision exercises will prove to be of particular value.

There is a school of thought which argues that all the major discrete distributions should be introduced before the continuous distributions. I have not followed this in the belief that many students too readily forget the early sections unless they are reinforced by exercises and subject material. Thus I have introduced the Poisson distribution in Chapter 11, though, much of the material can be introduced earlier if desired.

There is also a variety of approaches to the study of Probability. Teachers themselves have different preferences and students enter the course with differing sets of experiences. Some students may never have met the concept before; others will have been introduced, even at O level, to the elementary rules of Probability.

I do not believe we can study Probability at the formal, even if elementary level, without a firm foundation built at the intuitive level, so I have provided two chapters, separated by a chapter on measures of spread for variety. Chapter 3 assumes no previous knowledge and concentrates on developing a variety of skills, thus broadening the range of problems which can be tackled. Chapter 5 then covers the ground again more formally. An alternative approach is to begin with Chapter 5 and make excursions into the earlier chapters as appropriate. In this way I hope the development of this important subject will prove to be acceptable to all who use this book.

With the increasing and welcome use of calculators, the use of coding in calculations is frequently called into question. After much thought I have decided to retain its use for two reasons. Firstly: I believe we have a duty as teachers to ensure that all students who can solve a problem using a calculator should be able to solve it competently and reasonably efficiently without such use. Secondly: coding can be used to gain insight into the comparable theoretical results in Expectation Algebra. This is important because of the increasing emphasis placed on the introduction of Expectation Algebra, which is pleasing, as this allows for greater coherence in the development of proofs for the results stated in the text. On this point I have attempted to give proofs whenever possible. But it has to be accepted that many of the theorems are deep and difficult to prove using elementary mathematics, though the opportunity should not be lost of checking these results empirically.

I would like to thank the following examination authorities for permission to produce past examination questions: The Associated Examining Board, Oxford Delegacy of Local Examinations, Oxford and Cambridge Schools Examination Board (also for Mathematics for Education and Industry syllabus questions), University of Cambridge Local Examinations Syndicate, The Joint Matriculation Board (also for permission to include the statistical tables in the appendices) and The University of London University Entrance and School Examinations Council.

I would also like to mention my indebtedness to Ena Johnson and Gavin Wakefield for supplying some solutions to the exercises though, I am of course, responsible for the accuracy of these solutions.

Finally, I would like to thank Vince Driver for preparing the artwork and all the staff of University Tutorial Press for their assistance in the production of this book.

D. A. Bryars.

Introduction

A study of Statistics helps in making judgements when there is insufficient information to be certain of what will happen. For example, we may be interested in which party will win the next election. We cannot be certain of the result before voting takes place even if we contact every voter in the country. People may change their minds at the last minute. However, through the use of statistical methods it can be shown that we get reasonably accurate results by contacting relatively few people. This is more practical since it saves both time and money.

In a real investigation we are often overwhelmed by data and usually the first task is to summarise them to allow the important or salient features to emerge. An obvious method is to display the information in a diagram. Alternatively, we may attempt to summarise them by the use of a few numbers or 'statistics' as they are often referred to. This is the area of the subject called Descriptive Statistics.

Before the subject can be developed further we need to study Probability. This allows us to develop statistical models of the world. These simplified structures enable us to predict what will happen in the future, not with certainty but with some degree of confidence. This area might be termed Deductive Statistics.

Eventually these methods are combined in the study of Sampling Theory and Significance Tests. This is the area called Inferential Statistics. As the name suggests it is here that we learn how to make inferences or judgements, using the information which has been collected.

To summarise, a typical statistical investigation consists of:

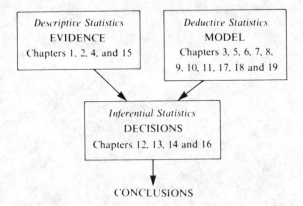

We begin with the 'evidence', the data collected for a statistical investigation. A clear distinction must be made between the various sorts of data. The most common data are numerical or *quantitative data.* The score obtained when a die is tossed, for example, or the height of certain people. *Qualitative data* are not numerical; for instance, the rank held by individuals in the armed forces or the type of tree found in a forest. One can often convert these sorts of data into *quantitative data* by coding. For example, when tossing a coin, code 'Head' as 1 and 'Tail' as 0. We shall not be particularly concerned with qualitative data. However, quantitative data further divide into discrete and continuous data. *Discrete data* can be established exactly and are found by *counting*. For example, the amount of money in a bank account, or the number of children in a family. *Continuous data* are obtained by measuring and cannot be established exactly, but in principle can be found to any degree of accuracy. For example, the length of a rod or the time taken to perform a certain task.

List of symbols

f	usually reserved for the frequency with which a particular value x occurs in a sample	22
Σ	'the sum of', e.g. Σx^2. More formally, $\displaystyle\sum_{i=1}^{n} x_i^2$	23, 24
$\bar{x}$	the mean of the x values, i.e. $\bar{x}=\dfrac{\Sigma fx}{\Sigma f}$	23
$P(\ldots)$	the probability of the event described in the bracket, e.g. $P(A)$, where A is a previously defined event	37
$n!$	n factorial	44
$^{n}P_{r}$	$\dfrac{n!}{(n-r)!}$	48
$^{n}C_{r}$	$\equiv\dbinom{n}{r}\equiv\dfrac{n!}{(n-r)!r!}$	49
Var (X)	the variance of X	62, 93
S_x	the sample standard deviation	63
Q_i	quartiles, $i=1, 2, 3$	72
$E, F, \ldots$	events	74
S	sample space	75
$P(E|F)$	conditional probability	77
X, Z	random variables, variates; x, z are particular values	91
$E(X)$	expectation of X	91
$B(n, p)$	Binomial distribution with parameters n and p	101
$f(x), f_X(x)$	(usually) a probability density function, pdf, of the variate X	117
μ, σ^2	mean and variance of a random variable	129, 173, 188
$\phi(x)$	pdf of the Normal distribution	129
$N(\mu, \sigma^2)$	Normal distribution with parameters μ (mean) and σ^2 (variance)	131
$\Phi(z)$	the Normal distribution function $\Phi(z)=P(Z\leqslant z)$ in N(0, 1)	131
$G(t)$	probability generating function	149
Cov (X, Y)	covariance of X and Y	152
$P(a)$	(in context) Poisson distribution with parameter a (mean)	157
S^2	sample variance	189
s^2	estimate of population variance	190
NH, AH	null and alternative hypothesis	201
χ^2	the χ^2 distribution (pronounced ki squared)	237

Chapter 1   Display of numerical data

1.1 Frequency distributions

Numerical data, whether obtained from surveys, experiments or a series of measurements, often need to be condensed into a more suitable form before they are of much value in a statistical investigation. Consider, for example, the marks gained by some students in a Mathematics test.

$$
\begin{array}{cccccccccc}
3 & 5 & 9 & 6 & 4 & 7 & 8 & 6 & 2 & 5 \\
10 & 1 & 6 & 3 & 6 & 5 & 4 & 7 & 8 & 4 \\
5 & 9 & 4 & 2 & 7 & 6 & 1 & 3 & 5 & 6 \\
2 & 6 & 4 & 8 & 3 & 1 & 7 & 9 & 7 & 2
\end{array}
$$

Data in this form are called *raw data,* and there are no obvious comments to make about these data, except that the highest mark is 10 and the lowest is 1. The difference between the highest and lowest scores, $10 - 1 = 9$, is called the *range.*

The problems with information given in this form is that no pattern is evident, nor are the data arranged so that simple facts, such as the number of students gaining 5 marks or more can be ascertained with ease. It is for these reasons that raw data are arranged into a *frequency distribution.*

Score	Tally chart	Frequency
1	|||	3
2	||||	4
3	||||	4
4	++++	5
5	++++	5
6	++++ ||	7
7	++++	5
8	|||	3
9	|||	3
10	|	1
	Total	40

Fig. 1.1 A frequency distribution to show the marks gained by 40 students in a mathematics test.

In the frequency distribution Fig. 1.1, the first column represents all the different possible scores and the frequency column represents the number of times each score occurs. Thus, for example, a score of 6 was gained by 7 students. Naturally the sum of the entries in the frequency column gives the total number of items of data, and can be used as a check when constructing a frequency distribution. Great care needs to be exercised since it is very easy to make an error. A good method is to form the central part of the table (the tally chart) step by step as each item of data is read off. Each fifth entry is marked ⊮⊬ rather than |||||| as it is easier to count in multiples of five. Frequency distributions often omit the central part of the table even though it is important in its construction.

It should be clear that even though we are considering a relatively small amount of data, a frequency distribution simplifies the information sufficiently to make it possible to notice the main features. In this distribution most of the marks are in the middle range, with fewer high or very low marks. This is typical of the distribution of marks in an examination, though the pattern was not evident from an inspection of the raw data.

1.2 Grouped frequency distributions

In many situations there are rather more data than those considered in the previous section. For example, suppose the data below represent the masses (to the nearest kg) of a group of youngsters.

```
70  38  54  47  68  51  34  45  65  82
67  22  73  56  74  35  85  52  41  38
62  57  75  53  44  60  68  59  50  28
38  40  64  45  77  52  53  47  56  72
51  80  54  61  69  52  35  79  68  57
32  51  33  42  73  26  84  40  56  74
55  45  67  61  58  49  37  60  76  52
61  34  55  82  36  78  50  48  62  60
53  65  74  41  62  45  21  76  35  59
65  88  35  48  63  70  46  55  43  36
```

It can be established by inspection that the greatest mass is 88 kg and the smallest is 21 kg, so the range is 67 kg. A frequency distribution of these data would be very long (having 68 rows!) and would not offer much simplification. We can, at the expense of losing information, group the data into the following *grouped frequency distribution*.

Class intervals (kg)	20–29	30–39	40–49	50–59	60–69	70–79	80–89
Frequency	4	14	17	25	20	14	6

The intervals 20–29, 30–39 etc. are called *class intervals*. The choice of intervals is essentially arbitrary, though for reasons which will become apparent, there should be between 5 and 15 such intervals (cf. Exercise 1.1 question 6). They should be so chosen as to allow each item of data to be allotted to the appropriate class interval without ambiguity. Thereafter the construction of the table is similar to that of the frequency distribution.

There are several problems associated with grouping data. Each of the above masses were given to the nearest kg, so a true mass of 49.4 kg would be rounded down to 49 kg, though for a mass of 49.5 kg a convention must be adopted. Thus any true mass between 39.5 and 49.5 kg (possibly including one or other of the end points, depending on the convention) would be counted as belonging to the class interval 40–49. The numbers 40 and 49 are called the *class limits* of the class interval 40–49 (and are usually given to the same degree of accuracy as the raw data). The numbers 39.5 and 49.5 are called the *class boundaries* of the class interval 40–49 (and will usually be given correct to one more significant figure than the raw data).

The difference between the lower and upper class boundaries is called the *class width* of any particular class interval and the value mid-way between the *class boundaries* is called the *mid-interval value* or *class mark*. Thus in our example, each class width is 10 and the mid-interval values are 24.5 34.5 44.5 54.5 64.5 74.5 and 84.5 respectively. There is no reason why all the class widths should be the same, since this depends on the choice of class intervals, though when possible it is usually convenient to make them so.

Suppose, however, that the raw data above referred to the ages in completed years of a group of adults. In this case, whilst the same class intervals may have been chosen, so the class limits remain the same, the class boundaries are different. The class boundaries of the class interval 40–49 would now be 40 and 50 since anyone a day younger than 40 years of age would be considered to be 39 years old and similarly anyone a day younger than 50 years would be considered to be 49 years old. The class width would still be 10, but the mid-interval value would be 45 in this case.

So far, in this section, we have considered *continuous data*, that is data which, in principle, could take any value in a certain range. Suppose now the raw data above represent the marks in an examination. This is an example of *discrete data*, which can only take specific values in a certain range; in this case the integers (presumably between 0 and 100). This time, supposing the same class intervals are adopted, we shall still define the class boundaries of, say, the class 40–49 as being 39.5 and 49.5 and thus the class width as 10 and the mid-interval value as 44.5, even though the boundaries no longer have the same interpretation as when dealing with continuous data. Thus, in all situations, the use of class boundaries enables us to cover the whole number line within the range, without leaving any gaps. The reason for these definitions and the care taken in their interpretation will become apparent in the next section, when histograms are dealt with.

Example 1.1

Consider the following grouped frequency distribution.

Class intervals	0–2	3–5	6–15	16 or more
Frequency	3	15	18	7

Write down the class boundaries, class widths and mid-interval value for each class interval in the case that the data represent:

(a) the ages in completed years of some children;
(b) the marks gained by some students in a test;
(c) the length to the nearest cm of some lines.

1.3 Methods of display

(a)	Class boundaries	0,3	3,6	6,16	16,18*
	Class widths	3	3	10	2
	Mid-interval values	$1\frac{1}{2}$	$4\frac{1}{2}$	11	17

*People who are 18 years old or more are adults.

(b)	Class boundaries	−0.5,2.5	2.5,5.5	5.5,15.5	15.5,20.5*
	Class widths	3	3	10	5
	Mid-interval values	1	4	10.5	18

*For the last class interval a guess was made that the test was out of 20. Notice also that since we are dealing with discrete data the first class boundary is −0.5.

(c)	Class boundaries	0,2.5	2.5,5.5	5.5,15.5	not known*
	Class widths	2.5	3	10	,,
	Mid-interval values	1.25	4.5	10.5	,,

*In view of the nature of the data we cannot really guess the class boundaries for the last class interval. For the first class interval we can take account of the fact that we cannot have a negative length since we are dealing with continuous data. This argument does not apply to the discrete data in part (b).

Frequency distributions

A convenient way of summarising data is in the form of a diagram, where the salient features can be ascertained at a glance. The more common methods of display are illustrated in this section, but the reader is recommended to take note of the many and varied methods employed in magazines and newspapers.

Consider the table introduced in 1.1 representing the marks gained by 40 pupils in a Mathematics test, reproduced below.

Mark	1	2	3	4	5	6	7	8	9	10
Frequency	3	4	4	5	5	7	5	3	3	1

The most appropriate diagram is the *line graph*, Fig. 1.2.

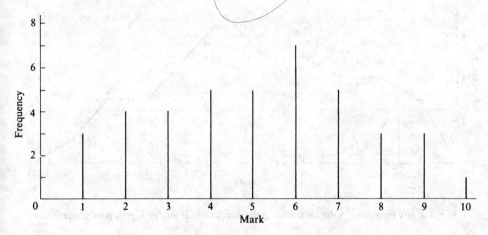

Fig. 1.2   A *line graph* to show the marks gained by 40 pupils.

The height of each line is proportional to the frequency with which the corresponding mark occurs.

Another method of display is the *histogram*, Fig. 1.3.

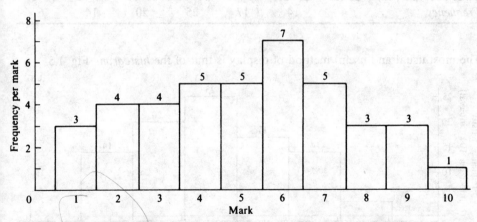

Fig. 1.3   A *histogram* to show the marks obtained by a number of students in a mathematical exam.

Another alternative is the *frequency polygon*, Fig. 1.4. A frequency polygon is obtained by joining the mid-points of the upper edges of the rectangles in the corresponding histogram by straight line segments. The corresponding histogram is shown super-imposed in Fig. 1.4 but is not normally present. Note that the graph has been extended at each end, which is usual whenever such an extension makes sense. An advantage of the frequency polygon is that two or more sets of data can be represented on the same diagram for purposes of comparison.

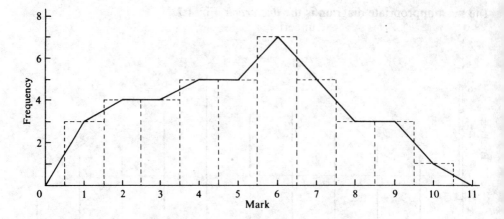

Fig. 1.4   A *frequency polygon* showing the marks obtained by a group of students.

Grouped frequency distributions

Consider the grouped frequency distribution representing the masses of 100 youngsters introduced in Section 1.2.

Class intervals (kg)	20–29	30–39	40–49	50–59	60–69	70–79	80–89
Frequency	4	14	17	25	20	14	6

The most usual and useful method of display is that of the *histogram*, Fig. 1.5.

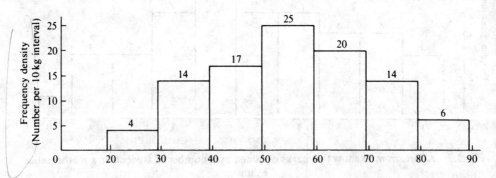

Fig. 1.5   A histogram showing the masses of 100 youngsters.

In a histogram, the horizontal axis is marked with a numerical scale (as with an ordinary graph) and above each class interval a rectangle is placed, extending to the *class boundaries,* such that the *area* of each rectangle is proportional to the frequency corresponding to that interval. In this example, the class widths are all the same so that the height of each rectangle is proportional to the frequencies. It is because the class widths may differ, however, that the vertical scale is called frequency density or in this example the frequency per 10 kg interval. As an alternative, a frequency polygon may be drawn. This can be derived from the histograms by joining the mid-points of the tops of the rectangles. In this case the underlying histogram is not shown.

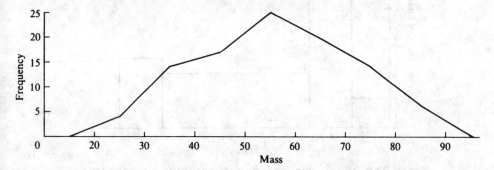

Fig. 1.6 A *frequency polygon* representing the same data as in Fig. 1.5.

Example 1.2 (cf. Example 1.1)

Draw a histogram to represent the data in the table below, when the data represent:

(a) ages in completed years of same children;
(b) the marks in a test;
(c) the length to the nearest cm of some lines.

Class intervals	0–2	3–5	6–15	16–18
Frequency	3	15	18	7

A discussion of the class boundaries for each of the above situations can be found in Example 1.1. See Fig. 1.7.

(a)

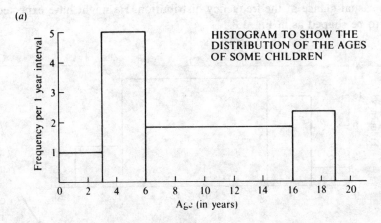

HISTOGRAM TO SHOW THE
DISTRIBUTION OF THE AGES
OF SOME CHILDREN

Fig. 1.7(a)

check

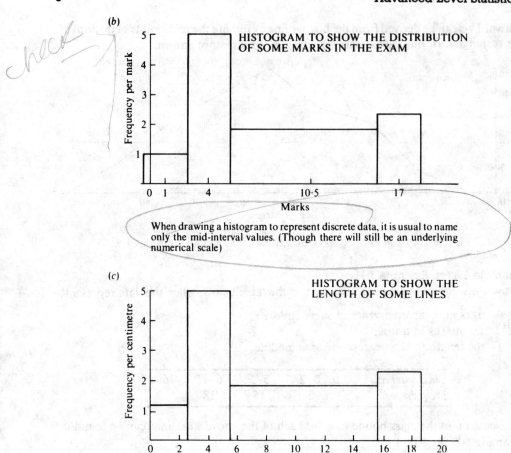

When drawing a histogram to represent discrete data, it is usual to name only the mid-interval values. (Though there will still be an underlying numerical scale)

Fig. 1.7(*b*) & (*c*)

The unwary reader may have been surprised by the shape of the histograms in Fig. 1.7, after a casual glance at the frequency distribution. He might have expected the histogram to be shaped as in Fig. 1.8.

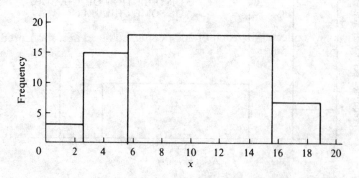

Fig. 1.8

Further thought and a consideration of the width of the class intervals will assure such a reader that the diagram above gives a completely false impression of the data.

Again consider the distribution of ages of some young children as represented in the following two frequency distributions.

Age (yrs)	1	2	3
Frequency	4	4	4

Table A

Age (yrs)	1	2 or 3
Frequency	4	8

Table B

Notice that these frequency distributions represent the same data.

The histograms for these frequency distributions are as follows:

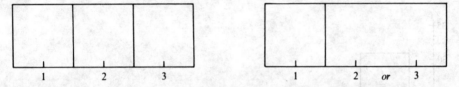

Both these histograms leave one with the same impression of the distribution of the ages. A completely false impression is given with the following diagram representing the data in Table B:

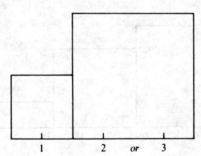

It is for this reason that the area and not the height, of each rectangle is proportional to the frequency.

Example 1.3

The amount of money collected by each of 300 people, involved in a sponsored walk, was recorded as follows:

Amount collected (to the nearest £)	1	2–4	5–10	11–15	16–25
Number of Children	20	93	90	58	39

Draw a histogram to represent these data.

The class boundaries are:

$\frac{1}{2}$, $1\frac{1}{2}$; $1\frac{1}{2}$, $4\frac{1}{2}$; $4\frac{1}{2}$, $10\frac{1}{2}$; $10\frac{1}{2}$, $15\frac{1}{2}$; $15\frac{1}{2}$, $25\frac{1}{2}$;

so the class widths are

1,   3,   6,   5,   10.

Dividing these into the corresponding frequencies gives the height of each 'rectangle' in the histogram

$$\frac{20}{1} = 20, \quad \frac{93}{3} = 31, \quad \frac{90}{6} = 15, \quad \frac{58}{5} = 11.6, \quad \frac{39}{10} = 3.9.$$

The histogram has been drawn in Fig. 1.9.

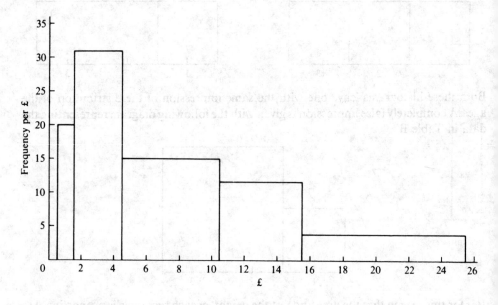

Fig. 1.9

If the vertical scale had been 'frequency per 50p' the height of each 'rectangle' would have been halved, and a choice of 'frequency per £5' would have increased each height by a factor of 5.

The most useful fact to emerge from a study of a diagram is its general shape. There are some distributions which occur so often that they have special names.

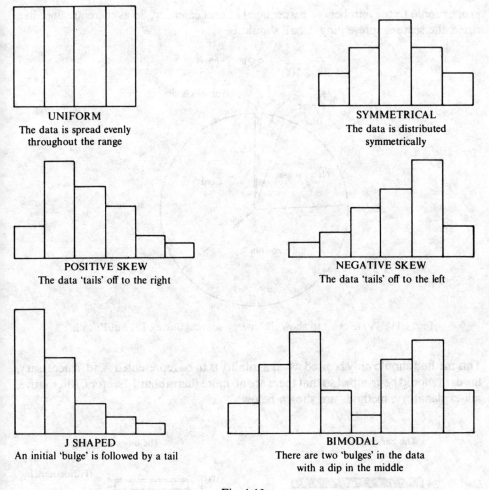

UNIFORM
The data is spread evenly
throughout the range

SYMMETRICAL
The data is distributed
symmetrically

POSITIVE SKEW
The data 'tails' off to the right

NEGATIVE SKEW
The data 'tails' off to the left

J SHAPED
An initial 'bulge' is followed by a tail

BIMODAL
There are two 'bulges' in the data
with a dip in the middle

Fig. 1.10

Finally, we consider the display of a common type of table where a non-numerical classification is given.

During 1977 our energy sources were approximately as given in the following table:

	%
Coal	36
Petroleum	40
Natural gas	19
Nuclear energy	4
Hydro electricity	1

To display information in this form the most appropriate diagram is a *Pie Chart*.

Here a circle (or pie) is divided into sectors (slices of pie) the area of each is

proportional to the number (or percentage) in that category. So as there are 360° in a
circle, the sector representing 'Coal' should be

$$\frac{36}{100} \times 360° \simeq 130°$$

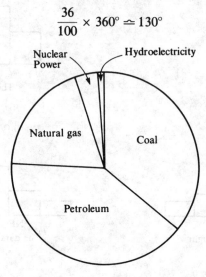

Fig. 1.11   A *pie chart* to show the energy sources during 1977 in Britain.

This method should only be used when a totality is to be represented; and, if necessary,
the data should be grouped so that there are no more than about 10 sectors. Alternative,
self-explanatory, methods are shown below:

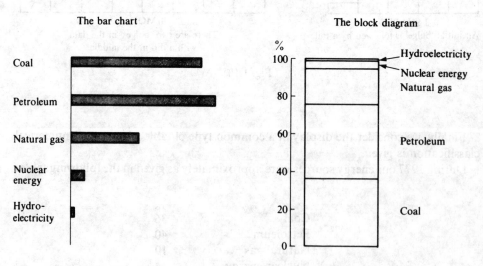

Fig. 1.12

Exercise 1.1

1. The following are the number of fires reported during 60 consecutive days in a certain town.

$$
\begin{array}{cccccccccccc}
4 & 2 & 2 & 4 & 1 & 1 & 0 & 1 & 0 & 0 & 3 & 2 \\
0 & 1 & 2 & 3 & 1 & 0 & 3 & 5 & 4 & 2 & 2 & 1 \\
3 & 2 & 6 & 1 & 0 & 4 & 0 & 4 & 3 & 4 & 0 & 1 \\
5 & 0 & 1 & 2 & 3 & 0 & 5 & 1 & 0 & 0 & 3 & 6 \\
1 & 2 & 5 & 4 & 0 & 2 & 0 & 5 & 1 & 3 & 0 & 2 \\
\end{array}
$$

Group these data into a frequency distribution and display the information.

2. The number of telephone calls received at an exchange in 50 successive 30 second intervals were as shown below.

$$
\begin{array}{cccccccccc}
5 & 3 & 4 & 2 & 1 & 3 & 3 & 6 & 6 & 3 \\
7 & 6 & 6 & 4 & 2 & 2 & 4 & 4 & 3 & 4 \\
7 & 4 & 3 & 6 & 5 & 6 & 4 & 2 & 5 & 3 \\
5 & 6 & 1 & 3 & 0 & 4 & 6 & 3 & 6 & 4 \\
1 & 3 & 1 & 7 & 4 & 6 & 3 & 2 & 3 & 5 \\
\end{array}
$$

Construct a frequency distribution from these data and draw a frequency polygon.

3. The height of members of a certain species of plant were recorded to the nearest mm, and grouped into a table having class intervals 10.5–11.4, 11.5–12.4, 12.5–13.5 cm.
Find the   (i) class boundaries,
          (ii) mid-interval values,
          (iii) class lengths.

4. The following are measurements of the breaking strength (in g) of a sample of 50 threads.

$$
\begin{array}{cccccccccc}
562 & 650 & 304 & 708 & 426 & 568 & 538 & 692 & 596 & 480 \\
596 & 490 & 620 & 424 & 566 & 542 & 500 & 654 & 506 & 668 \\
678 & 604 & 578 & 460 & 528 & 544 & 674 & 588 & 438 & 462 \\
493 & 591 & 389 & 580 & 784 & 584 & 470 & 493 & 487 & 690 \\
721 & 743 & 670 & 582 & 472 & 574 & 696 & 584 & 472 & 518 \\
\end{array}
$$

Group these measurements into a distribution having class intervals 300–349, 350–399 ... 750–799 and construct a histogram.

5. A sample of 100 capacitors, each of nominal capacitance $4\,\mu F$, were measured accurately and found to have a distribution as shown below.

Capacitance (Mid-interval value)	3.85	3.90	3.95	4.00	4.05	4.10	4.15
Frequency	1	13	21	29	22	11	3

Display this information by means of a histogram.

6. The marks gained by some students in an examination were as shown below. Group the data and display the information by a suitable histogram. Describe the general shape and suggest reasons for obtaining such a shape.

65	12	1	21	34	25	61	17	45	58	73
19	50	2	24	71	2	68	32	16	49	16
41	11	52	64	22	50	14	6	26	70	59
18	56	67	13	5	17	77	63	57	9	75
15	21	52	29	13	67	37	25	46	57	29
17	84	5	42	67	43	14	7	57	15	69
48	24	69	23	39	53	64	64	65	28	55

Note: It is worthwhile experimenting with the number of class intervals in order to justify the claim made in Section 1.2 that between 5 and 15 class intervals is appropriate. This claim was made in anticipation of drawing histograms.

7. The following data are the times taken to travel to work each day (in hours) for workers at a certain factory.

Under $\frac{1}{2}$ hour	58
$\frac{1}{2}$ hour —	35
1 hour —	6
$1\frac{1}{2}$ hours —	3
2 hours —	2
$2\frac{1}{2}$ hours–3 hours	1

The notation '1 hour —' means 1 hour or more, but less than $1\frac{1}{2}$ hours.

Draw a histogram to represent this information. Do you think the shape is typical for the workers at a factory? Would the same shape occur for the times taken to travel to your school (perhaps measured in 10 minute intervals)?

8. Two similar types of components, A and B were tested to failure. The results are summarised below.

Time to failure (in hrs)	less than 20	20–39	40–59	60–79	80–99	100–139
Type A	8	8	7	7	4	16
Type B	1	2	7	20	15	5

Compare these two distributions graphically. Which type of components would you prefer to use?

9. The table below shows the typical values of the mass of a human being during the first year of life.

Age (months)	Birth	1	2	3	4	5	6	7	8	9	10	11	12
Weight (kg)	2.9	3.9	5.0	6.0	6.7	7.4	8.0	8.4	8.8	9.1	9.4	9.7	9.9

Display this information using a suitable diagram and estimate the mass after six weeks.

Produce a new table to indicate the gain in weight per month and display this data in the form of a bar chart. Compare and contrast the different impressions given by each of your diagrams.

10. The array below gives the intelligence quotient (IQ) of 100 children. Group the data into class intervals 65–74, 75–84 ... 125–134 and draw a histogram to illustrate the data.

		units									
		0	1	2	3	4	5	6	7	8	9
	60						1				
	70	1			2		1		1	1	
	80		1		1	3		1	3	1	2
Tens	90	1	2	2	3	2	2	4	1	5	7
	100	5	4	5	3	6	2	3	2	2	1
	110	2	3	1	2	1	1		1	2	
	120		1		1	1				1	
	130	1				1					

11. The table below shows some leisure activities for different age groups. Display the information by histograms. Describe their shapes and salient features.

	% persons engaged in activity frequently			
	Age			
	18–29	30–44	45–59	60–65
Watching TV	93	91	92	91
Gardening	7	22	26	35
Taking part in sport	34	23	13	8
Going to bingo	6	13	16	8

12. The age distribution of the population of England, Wales and Scotland was recently found to be as follows (in 1000's).

Under 5	4349
5 —	4512
10 —	4069
15 —	3705
20 —	4121
25 —	6676
35 —	6333
45 —	6651
55 —	6414
65 —	4604
75 —	2072
85 and over	461
Total	53 967

Display this data in the form of a histogram.

13. An attempt was made to ascertain the main mode of transport to and from work, for people living in areas of differing population densities.

	Rail	Bus	Car/Taxi	Other (Walking, Bicycle, etc.)
Large Conurbation (London)	18	18	39	25
Urban Area	2	22	48	28
Rural	1	11	63	25

The figures represent percentages.

Display this information in any way you feel is appropriate and discuss the significance of the data. Comment on any difficulties you would envisage in producing such a table.

14. The Broadcast Receiving Licences current at the end of each month (in 1000s) in 1978 were as shown below.

	Dec '77	Jan '78	Feb	March	April	May
Black and White	7367	7252	7182	7100	7052	7008
Colour	10 721	10 927	11 002	11 049	11 148	11 257

	June	July	Aug	Sept	Oct	Nov	Dec
Black and White	6967	6909	6856	6772	6699	6600	6519
Colour	11 328	11 356	11 469	11 519	11 655	11 799	11 973

Display this information in the same diagram and comment on the salient features.

15. During 1974 our energy sources were approximately as given in the following table.

	%
Coal	35
Petroleum	45
Natural gas	14
Nuclear energy	4
Hydro electricity	1

In Section 1.3. the equivalent information was quoted for 1977. Display both sets of data on the same diagram and discuss the main features of the data. Does your diagram suggest that the amount of petroleum used has increased in the three year period? Is this necessarily true?

16. During the tax year 1977/78 the Inland Revenue received the following (in £1 million)

Income Tax	17 420
Corporation Tax	3346

Capital Gains Tax	340
Capital Transfer Tax	312
Stamp duties	375
Other	124
Total	21 927

Draw a pie chart to represent this data. So, for example, the size of the sector representing income tax is proportional to the angle

$$\frac{17\,420}{21\,917} \times 360° \simeq 286°$$

17. During the 1983 general election the votes cast and the seats won by the various parties were as follows:

	% Votes	Seats won
Conservative	43.5	397
Labour	28.3	209
Alliance	26.0	23
Others	2.2	21
	100	650

(a) Draw pie charts to represent the distribution of (i) the votes, (ii) the seats won during this election.

(b) Calculate the number of seats which would have been won by the various parties had they been allocated on a percentage basis. Compare these with the actual seats won by drawing a suitable diagram. Comment on the features the diagram illustrates. Can you explain the discrepancy?

18. The expectation of life (in years) at various ages, by sex, for two particular periods is shown in the following table.

	Expectation of life (in years)			
	1870		1970	
Age	Male	Female	Male	Female
0	40.5	42.8	68.5	75
15	44.0	44.9	55.6	61.5
25	36.8	38.0	46.1	58
45	23.1	24.4	27.3	32.6
65	10.9	11.8	12.2	15.8

Study this information and explain what is meant by the number *61.5* in italics above.

Display the data as clearly as possible and describe the main features of the information.

19. During the 1981 census it was established that there were 1 303 948 residents of South Yorkshire. The percentage of the population in each age group was as follows:

Age (years)	· 0–4	5–15	16–24	25–34	35–44	45 —
%	5.8	16.7	14.3	13.9	11.9	37.3

Display this information in the form of a histogram. (Assume the upper age limit to be 90 years)

20. The table below shows the percentage of earnings taken in direct tax for certain groups of workers over a period of time.

	Low paid	Average earners	High paid
1971–72	21	19	16
1972–73	20	18	15
1973–74	21	19	16
1974–75	24	22	19
1975–76	27	24	22
1976–77	27	24	22
1977–78	24	22	19
1978–79	23	21	18
1979–80	21	19	17
1980–81	22	20	17

Display this information in a suitable form and comment on the salient features.

Chapter 2   Averages

2.1 The mean, mode and median

The preceding chapter has shown how raw data can be reduced by means of tables, and the salient features can be portrayed by means of diagrams. The next step is to attempt to summarise the information by means of a single number which serves as a representative of the data. Such a number is called an average or, since it is usually close to the middle of the data when arranged according to size, it is also known as a *measure of central tendency*. During the course of this chapter we shall develop relatively sophisticated methods for calculating these averages as the amount of data under consideration is increased.

Suppose John gains the following marks in his school examination:

$$48, \quad 64, \quad 55, \quad 64, \quad 74$$

For convenience and interest he will calculate an 'average' mark, a number which is typical of the marks above. There are three ways to obtain such a number in general use.

The mean
The mean of a set of raw data is obtained by adding together all the items and dividing by the number of items.

Thus John's mean score is

$$\frac{48 + 64 + 55 + 64 + 74}{5} = \frac{305}{5}$$

$$= 61$$

The median
The median is defined as the middle number when the data is arranged in order of size. If there is an even number of items and hence two 'middle' numbers then the median is the value mid-way between these two.

Again, in John's case the median would be 64:

$$48, \quad 55, \quad 64, \quad 64, \quad 74$$

The mode
The mode is defined as the number which occurs most frequently (the most 'popular' number). The mode may not exist, because too many numbers occur equally frequently. There may, however, be two modes. In John's case the mode is 64.

There is no 'correct' average as each has its uses. An important property of the mean is that it involves all of the data in its calculation. This can be an advantage as it is sensitive to small changes in the data.

Though, likewise, it can be a disadvantage as it is strongly affected by extreme values.

Suppose, for example, that Peter scores

$$22, \quad 26, \quad 19, \quad 23, \quad 85$$

in the same examination, the averages for this data are

> Mean   35,
> Median   23,
> Mode   There is no mode in this case.

One could hardly claim that the mean is a fair representative of this data, whereas the median is much more reasonable.

This highlights a particular property of the median in that it is not affected by extremes, since it is only concerned with the central values when the data is arranged according to size.

It would appear that the mode is of little interest, despite its ease of calculation, though shirt and shoe manufactures would be interested! In Statistics, however, the most important average is the mean, partly because of its sensitivity and partly because of the ease of dealing with it mathematically.

Example 2.1

In 12 consecutive innings during a season a batsman's scores were:

$$6, \quad 13, \quad 16, \quad 45, \quad 93, \quad 0, \quad 62, \quad 87, \quad 136, \quad 25, \quad 14, \quad 31$$

Find his mean score and the median.

Before the end of the season the batsman had 8 more innings and thereby decreased his mean score for the season by 4 runs. Find the mean score for the last 8 innings.

We rearrange the data according to size:

$$0, \quad 6, \quad 13, \quad 14, \quad 16, \quad 25, \quad 31, \quad 45, \quad 62, \quad 87, \quad 93, \quad 136$$

The median is mid-way between 25 and 31 which is

$$\frac{25 + 31}{2} = 28.$$

The mean score is

$$\frac{528}{12} = 44,$$

where the total for the first 12 innings is 528.

Over the whole 20 innings his mean score is reduced to 40. So

$$\frac{\text{'total score in 20 innings'}}{20} = 40$$

and hence his total score is 800, giving his total in the last 8 innings as

$$800 - 528 = 272.$$

So the mean for the last 8 innings is

$$\frac{272}{8} = 34.$$

Exercise 2.1

1. Find the mean, mode and median of the following:

$$3, \ 6, \ 0, \ 5, \ 8, \ 2, \ 5, \ 7, \ 4, \ 0.$$

2. (See text above) John's chemistry mark is 67%. Find his new mean, mode and median.

3. The mean of the following set of numbers is 6 and the mode is 5, find the median

$$7, \ 9, \ x, \ 7, \ y, \ z, \ 5, \ 20.$$

4. (i) Under what circumstances will the mean, mode and median coincide?
   (ii) Write down the shortest sequence of numbers you can so that

   (a) Mean < mode < median
   (b) Mode < median < mean

5. The following are the IQs of 12 people.

   115, 89, 94, 107, 98, 87, 99, 120, 100, 94, 100, 99.

   It is claimed that 'the average person in the group has an IQ of over 100'. Is this a reasonable assertion?

6. The typing speeds (number of words per minute) of 10 girls were

   37, 52, 41, 40, 43, 48, 44, 58, 39, 42.

   Calculate the mean typing speed and state the number of girls who attained a speed greater than the mean. Write down the number of girls who achieved a score greater than the median.

7. A sample consisting of 10 very expensive components was tested to destruction, to establish how long they would last. The times to failure, recorded in hours during the first three days of testing were

   40, 44, 55, 55, 64, 69.

   Which would be the most appropriate average to describe the life of these components? Discuss the consequences of your choice.

8. The weekly earnings (in £) of a group of workers in a factory are shown below.

Weekly earnings (£)	40	50	70	90
Number of men	5	40	12	3

Which average representing the weekly wage of the whole group of workers would alter if 10 employees worked overtime and increased their weekly wage by £10 each?

9. Find the mean of the following:

$$3, \quad 6, \quad 5, \quad 1, \quad 7.$$

Hence write down the mean of

(i) 13,   16,   15,   11,   17
(ii) -2,   1,   0, -4,   2
(iii) $a + 3$, $a + 6$, $a + 5$, $a + 1$, $a + 7$
(iv) 30,   60,   50,   10,   70
(v) $3b$, $6b$, $5b$, b, $7b$
(vi) $a + 3b$, $a + 6b$, $a + 5b$, $a + b$, $a + 7b$

10. Find the mean, mode and median of the following data:

$$1,1,1,1,1 \quad 2,2,2,2,2 \quad 3,3,3,3,3 \quad 4,4,4,4 \quad 5,5,5,5,5 \quad 6,6,6,6,6$$

11. Find the mean of the following set of numbers:

$$\text{six 5s,} \quad \text{four 10s,} \quad \text{two 15s} \quad \text{and} \quad \text{one 20}$$

12. The average salary paid to the managers in three companies is £7000, £6000 and £9000 p.a. respectively. If the respective number of managers in these companies is 5, 12, 3 find the mean salary paid to the 20 managers.

2.2 Averages from frequency distributions

It was noted in Section 1.1 that when dealing with a lot of raw data it is often convenient to arrange them into a frequency distribution. In this section we consider methods of obtaining averages when the data is so arranged. Suppose the following data represent the scores obtained when a die was tossed sixty times:

x	1	2	3	4	5	6	Score
f	11	7	8	10	13	11	Frequency

To compute the *mean* we need to add all the scores obtained and divide by 60. Now a 'one' occurred 11 times which contributed 11 to the sum and a 'two' occurred 7 times contributing 14 to the sum and so on. That is

$$\text{mean} = \frac{(11 \times 1) + (7 \times 2) + (8 \times 3) + (10 \times 4) + (13 \times 5) + (11 \times 6)}{60}$$

More generally if x can take values

$$x_1, x_2, x_3, x_4, \ldots, x_m,$$

with corresponding frequencies

$$f_1, f_2, f_3, f_4, \ldots, f_m,$$

then

$$\text{mean} = \frac{f_1 x_1 + f_2 x_2 + f_3 x_3 + \ldots + f_n x_n}{f_1 + f_2 + f_3 + \ldots + f_n}.$$

The symbol Σ (pronounced Sigma) is used to indicate 'the sum of'.

Thus Σf means the sum of the frequencies and Σfx means the sum of the products fx; that is

$$\Sigma f = f_1 + f_2 + f_3 + \ldots + f_n$$

and

$$\Sigma fx = f_1 x_1 + f_2 x_2 + f_3 x_3 + \ldots + f_n x_n$$

The notation $\bar{x}$ will be used to indicate the mean of the x values. (Similarly $\bar{u}$ is used to indicate the mean of a set of u values).

Combining these definitions we have:

Definition 2.1

$$\text{Mean} = \bar{x} = \frac{\Sigma fx}{\Sigma f}$$

The following indicates a method of setting out the calculations for finding the mean $\bar{x}$ using the above data.

x	f	fx*
1	11	11
2	7	14
3	8	24
4	10	40
5	13	65
6	11	66
	$\Sigma f = 60$	$\Sigma fx = 220$

*Values in this column are obtained by multiplying the corresponding values in the other two.

$$\bar{x} = \frac{\Sigma fx}{\Sigma f}$$

$$= \frac{220}{60}$$

$$= 3.67$$

A calculator can be used to good effect here, particularly if it possesses a memory. In this case the 'partial sums' in the 'fx' column can be stored directly in the memory, and the sum Σfx extracted at the end. However, particular care must be taken as errors can be difficult to trace.

The *mode* can be found by inspection of the frequency distribution. In this case a 'five' occurs most frequently.

To find the *median* note that there are 60 values in this case, so there are two middle numbers, the 30th and 31st. It can be seen from the frequency distribution that both the 30th and 31st numbers (when arranged in order of size) are equal to 4.
Thus the median is 4.

Definition 2.2

More generally if there are N numbers, where N is odd, then the unique middle number is the $\frac{1}{2}(N + 1)^{\text{th}}$ when ordered by size. If N is even then the two middle numbers are the $\frac{1}{2}N^{\text{th}}$ and $(\frac{1}{2}N + 1)^{\text{th}}$.

2.3 Sigma notation

Suppose x can take the values $x_1, x_2, x_3, \ldots, x_n$ then we write the sum

$$x_1 + x_2 + x_3 + \ldots + x_n = \Sigma x$$

A more precise notation would be $\displaystyle\sum_{i=1}^{n} x_i$

so

$$\sum_{i=1}^{4} x_i = x_1 + x_2 + x_3 + x_4$$

and

$$\sum_{i=1}^{5} i^2 = 1^2 + 2^2 + 3^2 + 4^2 + 5^2$$

In this book there will be no need to use such precise notation since the context will always allow us to expand the expression.

However, care does need to be exercised when dealing with summations.

Example 2.2

If x can take values $x_1, x_2, x_3, x_4, \ldots, x_n$ with associated frequencies $f_1, f_2, f_3, f_4, \ldots, f_n$, write out the following in full and simplify when possible.

(a) $\Sigma 3x$,
(b) Σx^2,
(c) Σfx^2,
(d) $\Sigma(fx + a)$
(e) $\Sigma(ax + b)$ } where a and b are constants.

(a) $\Sigma 3x = 3x_1 + 3x_2 + 3x_3 + \ldots + 3x_n$

$\qquad = 3(x_1 + x_2 + x_3 + \ldots + x_n)$

$\qquad = 3\Sigma x$

(b) $\Sigma x^2 = x_1^2 + x_2^2 + x_3^2 + \ldots + x_n^2$

$\qquad$ (There is no simplification in this case.)

(c) $\Sigma fx^2 = f_1 x_1^2 + f_2 x_2^2 + \ldots + f_n x_n^2$

(d) $\Sigma(fx + b) = (f_1 x_1 + b) + (f_2 x_2 + b) + \ldots + (f_n x_n + b)$

$\qquad\qquad = \Sigma fx + nb$

(e) $\Sigma(ax + b) = (ax_1 + b) + (ax_2 + b) + \ldots + (ax_n + b)$

$\qquad\qquad = a\Sigma x + nb$

Notice that if we take $a = 0$ in part (e) we obtain

$$\Sigma b = nb$$

and further, if we take instead $b = 0$ in part (e), we obtain

$$\Sigma ax = a\,\Sigma x \quad \text{where } a \text{ is a constant,}$$

However, Σfx cannot be so simplified since f varies with x.

In summary:

Statement 2.1

$$\Sigma ax = a\Sigma x$$

$$\Sigma b = nb \left(\text{more precisely } \sum_{i=1}^{n} b = nb \right)$$

$$\Sigma(ax + b) = a\Sigma x + nb$$

where a and b are constants and x can take n values.

2.4 Coding to find the mean

When we wish to find the mean of a lot of data even using a frequency distribution can result in tedious calculations. The work can be eased considerably if, instead of using the x column to find $\bar{x}$, we produce a new column, which we call u, where $x = M + Cu$ and we calculate $\bar{u}$. If we choose the constants M and C appropriately we find the arithmetic much easier to handle. The method is explained by way of an example.

Example 2.3

Find the mean of x in the following frequency distribution:

	x	f	fx	$u = \dfrac{x-27}{5}$	fu
	12	4	48	−3	−12
	17	9	153	−2	−18
	22	18	396	−1	−18
$M \rightarrow$	27	17	459	0	0
	32	12	384	1	12
	37	3	111	2	6
		$\Sigma f = 63$	$\Sigma fx = 1551$		$\Sigma fu = -30$

Direct approach   *Alternative approach*

$$\bar{x} = \frac{\Sigma fx}{\Sigma f}$$

(See below)

$$= \frac{1551}{63}$$

$$\simeq 24.6$$

Alternative approach: Method

Choose one of the x's (in our example we choose 27) and take this to be M (called the 'trial mean').

It does not matter which number we choose though it will soon be realised that the nearer the choice is to the actual mean the easier the resulting numbers are to deal with.

Choose C to be the common difference between the x's whenever there is one, in other cases take C to equal 1. In our example we take $C = 5$.

Form a u column by setting

$$u = \frac{x-27}{5}$$

and now compute $\bar{u}$ in the usual way

$$\bar{u} = \frac{\Sigma fu}{\Sigma f}$$

$$= \frac{-30}{63}$$

$$\simeq -0.48.$$

It will be shown below that since

$$x = M + Cu$$

$$\bar{x} = M + C\bar{u}$$

Thus

$$\bar{x} = 27 + (5 \times -0.48)$$

$$\simeq 24.6$$

In agreement with the original method

Alternative approach: Explanation

Suppose quite generally that x takes values

$$x_1, x_2, x_3, \ldots, x_n$$

with corresponding frequencies $f_1, f_2, f_3, \ldots, f_n$ and that u is related to x by $u = \dfrac{x - M}{C}$

that is $x = M + Cu$ where M and C are constants.

$$\bar{x} = \frac{\Sigma f x}{\Sigma f} = \frac{f_1 x_1 + f_2 x_2 + \ldots + f_n x_n}{\Sigma f}$$

$$= \frac{f_1(M + Cu_1) + f_2(M + Cu_2) + \ldots + f_n(M + Cu_n)}{\Sigma f}$$

$$= \frac{M(f_1 + f_2 + \ldots + f_n) + C(f_1 u_1 + f_2 u_2 + \ldots + f_n u_n)}{\Sigma f}$$

$$= \frac{M\Sigma f}{\Sigma f} + \frac{C\Sigma fu}{\Sigma f}$$

$$= M + C\bar{u}$$

Example 2.4

Find an approximate value of the mean weekly wage of a group of 100 workers whose earnings are indicated below.

Weekly wage (nearest £10)	70	80	90	100	110	x
Number of workers	25	13	31	21	4	f

	x	f	$u = \dfrac{x - 90}{10}$	fu
	70	25	−2	−50
	80	13	−1	−13 (−63)
$M \rightarrow$	90	31	0	0
	100	27	1	27
	110	4	2	8 (35)
		$\Sigma f = 100$		$\Sigma fu = -28$

If

$$u = \frac{x - 90}{10}$$

then

$$x = 90 + 10u$$

So

$$\bar{u} = \frac{\Sigma fu}{\Sigma f} = \frac{-28}{100} = -0.28$$

Since

$$x = 10u + 90$$

$$\bar{x} = 10\bar{u} + 90$$

$$= -2.8 + 90$$

$$= 87.2$$

Thus the approximate mean weekly salary is £87.00.

It would be misleading to give the answer as £87.20 since this degree of accuracy is not reflected in the accuracy of the data.

Exercise 2.2

1. Find, without coding, the mean of the following data.

x	3	7	11	14
f	5	4	3	8

2. Use coding to find the mean of the following distributions.

(a)

x	10	20	30	40	50
f	16	18	25	19	22

(b)

x	121	130	139	148	157	166	175
f	3	8	11	16	20	25	31

3. Form a frequency distribution from the following raw data and hence find the mean:
   (i) without using coding;
   (ii) by coding.

```
6 8 3 6 4 3 7 9 1 2 6 5 7 3 4 7 4 3 7 3
6 1 4 6 9 8 6 3 7 1 6 2 3 3 2 6 1 6 8 0
4 5 6 0 1 1 1 4 1 0 9 5 9 7 7 4 2 4 6 7
6 2 4 2 8 1 1 4 5 7 2 0 4 2 5 3 3 2 3 7
3 2 1 6 7 6 6 2 2 7 6 6 5 6 5 0 2 6 7 1
```

4. The data below represents the marks of 50 students in an examination, obtain the mean mark.

$$
\begin{array}{cccccccccc}
41 & 24 & 51 & 40 & 51 & 25 & 8 & 63 & 19 & 47 \\
30 & 59 & 22 & 41 & 3 & 49 & 39 & 34 & 39 & 53 \\
48 & 78 & 37 & 50 & 70 & 11 & 13 & 27 & 48 & 20 \\
54 & 18 & 60 & 33 & 18 & 27 & 30 & 17 & 69 & 89 \\
36 & 52 & 58 & 38 & 55 & 40 & 43 & 45 & 35 & 57
\end{array}
$$

5. Choose two books, one a modern novel and the other a classic. From each take 100 consecutive words and obtain the number of letters in each word. Form two frequency tables and hence find the mean word length for each of your two samples.

Now for each book consider 100 consecutive sentences and count the number of words in each sentence. Find the mean length of sentences in each of these books. Comment on your results, drawing any conclusions you feel are justified.

6. A shopkeeper who sells fireworks keeps a record of the price of the first 60 sold.

$$
\begin{array}{cccccccccc}
50, & 30, & 10, & 10, & 90, & 10, & 20, & 10, & 10, & 10, \\
60, & 30, & 20, & 30, & 60, & 90, & 20, & 40, & 20, & 40, \\
60, & 10, & 20, & 40, & 90, & 50, & 40, & 10, & 20, & 20, \\
10, & 40, & 50, & 30, & 30, & 90, & 90, & 60, & 10, & 90, \\
60, & 40, & 10, & 50, & 20, & 40, & 60, & 20, & 30, & 20, \\
10, & 10, & 90, & 60, & 20, & 10, & 10, & 20, & 20, & 10
\end{array}
$$

Draw a frequency table and find the mode. Which firework do you think he would least wish to keep selling? Calculate how much money he took for each type of firework. Comment on your result.

7. The following table gives the distribution of the number of beans per pod in two types of broad bean. Draw histograms for each distribution. Find the mode and the median number of beans per pod for each type of bean.

Number of beans per pod	1	2	3	4	5	6
Frequency of type A	17	29	36	16	0	0
Frequency of type B	4	28	41	35	4	3

Which bean would you plant for a high yield?

8. Find the mean of the following distribution by coding as

$$u = \frac{x - 1.41}{0.1}, \quad \text{so } x = 0.1\,u + 1.41$$

x	1.21	1.31	1.41	1.51	1.61
f	1	4	9	8	3

9. Use coding to find the mean of the following distributions.

(a) x 5.0   5.2   5.4   5.6   5.8   6.0
   f 3   8   11   17   7   4

(b) x 1000   5000   10 000   15 000   20 000
   f 6   12   14   7   1

(c) x 16.411   16.412   16.413   16.414   16.415
   f 1   9   5   3   2

2.5 Averages from grouped frequency distributions

If we wish to find 'averages' when data is arranged in the form of a grouped frequency distribution, we cannot expect to give exact answers since information has been lost in the process of grouping. In the distribution below we no longer know, for example, what the 19 numbers in the class interval 41–60 are.

Class interval	1–20	21–40	41–60	61–80	81–100
f	9	17	19	4	1

We might investigate the original raw data, but if this is no longer available a sensible idea would be to assume the numbers are evenly spaced in that interval, so that the 'average' of these numbers could be taken as 50.5 (the mid-interval value – cf. Section 1.2).

This enables us to estimate the mean, mode and median as follows:

The mean

To find the mean form an x column representing the mid-interval values of each class interval and then proceed as in the previous section.

Class interval	f	x	$u = \dfrac{x - 30.5}{20}$	fu
1–20	9	10.5	−1	−9
21–40	17	$M \to$ 30.5	0	0
41–60	19	50.5	1	19
61–80	4	70.5	2	8
81–100	1	90.5	3	3
	$\Sigma f = 50$			$\Sigma fu = 21$

Since

$$u = \frac{x - 30.5}{20}$$

$$x = 30.5 + 20u$$

So

$$\bar{u} = \frac{\Sigma fu}{\Sigma f} = \frac{21}{50}$$

Thus

$$\bar{x} = 30.5 + 20 \times \frac{21}{50}$$

$$\simeq 38.9$$

$$\simeq 39$$

This compares reasonably well with the mean obtained directly from the raw data, 40.28 (see Exercise 2.2, question 4), but shows that it is pointless to find the mean to many significant figures. Occasionally a distribution is presented with the length of one or both of the extreme class intervals undefined. In this case a mid-interval value has to be estimated, if this is possible. Once the mid-values have been found a calculator can be used directly, rather than code.

The mode

There is no method of finding the mode of these data. The best we can do is to give the modal class – the class containing most numbers, which in this case is 41–60.

The median

The median in this case would be the number mid-way between the 25th and 26th number, when the data are written out in order of size. We cannot find these numbers from a grouped frequency distribution, but can say that the median will be about the 16th number in the class interval 21–40 which contains 17 numbers, assumed to be evenly spaced. So the median will be approximately 40. An alternative method for estimating the median will be considered in the next section.

Exercise 2.3

1. On a certain stretch of road the speeds of passing cars were recorded as follows.

Speed (mph)	10	20	30	40	50	60–80
Number of cars	17	36	28	4	2	3

Find the mean and modal class of this distribution. What do you think was the speed limit on this road? Find the percentage of cars breaking this limit.

2. Form a frequency distribution of the following data with intervals centred at 10, 15, 20, 25, 30, 35, 40 and estimate the mean and median.

$$
\begin{array}{cccccccccc}
9 & 26 & 33 & 24 & 41 & 24 & 37 & 27 & 39 & 30 \\
28 & 34 & 19 & 32 & 24 & 42 & 17 & 26 & 18 & 33 \\
40 & 28 & 31 & 20 & 23 & 18 & 21 & 32 & 21 & 39 \\
25 & 16 & 17 & 26 & 11 & 30 & 28 & 24 & 27 & 40
\end{array}
$$

3. The frequency distribution of the length of a sample of 100 nails, measured to the nearest 0.1 mm is shown below.

Class intervals	f
4.0–4.2	4
4.3–4.5	9
4.6–4.8	13
4.9–5.1	20
5.2–5.4	34
5.5–5.7	18
5.8–6.0	2

Find the mean length of this sample.

4. The times of arrival of trains at a particular station were recorded, in minutes, relative to the times they were due. The results are shown below.

Time of arrival	−8	−4	−0	4	8	12	16–30
Frequency	70	65	20	10	10	15	10

Find the mean time of arrival, and explain what this figure means.

5. Find the mean of the following distribution.

Class interval	0.20–0.24	0.25–0.29	0.30–0.34	0.35–0.39
Frequency	6	12	19	13

6. Estimate the mean and median of the distribution given below.

Class interval	Frequency
16.50–16.59	25
16.60–16.69	47
16.70–16.79	65
16.80–16.89	47
16.90–16.99	16

2.6 Cumulative frequency polygons

The marks of 400 candidates in an A level examination are given in the grouped frequency table below (Fig. 2.2).

Marks (%)	f	Upper class boundary	Cumulative frequency
0–10	6	10.5	6
11–20	15	20.5	21
21–30	31	30.5	52*
31–40	80	40.5	132
41–50	93	50.5	225
51–60	69	60.5	294
61–70	54	70.5	348
71–80	33	80.5	381
81–90	12	90.5	393
91–100	7	100	400
	$\Sigma f = 400$		

*Indicates that 52 candidates gained less than 30.5%.

Fig. 2.2 An example of a cumulative frequency table (CF table) obtained from a frequency table.

Finding the mean and modal class from a grouped frequency table causes no difficulties now, but finding the median involves some calculations which have yet to be analysed.

The grouped frequency table is extended by first finding the upper class boundaries with as much care as shown previously (cf. Section 1.2). Then the cumulative frequency column (CF column) is added. This is formed at each point of the table by adding to the frequency associated with the corresponding class interval the sum of the frequencies associated with the previous class intervals. Thus the entry of 52 in the CF column indicates that there are 52 candidates with a mark of 30.5% or less. Naturally if the frequency table had referred to the age in completed years then the upper class boundaries would be different and the entry of 52 would have indicated that there were 52 people under 31 years of age.

To find the median we need to find the class interval which contains the 200th number. The CF column indicates that the 200th number, the median, is in the 41–50 class interval. Strictly in this case we should find the mark mid-way between the 200th and 201st mark, but such accuracy is pointless in a grouped frequency table since there is so much information missing.

Inspection of the figures in the frequency table indicates that the 200th number is approximately the 68th number in the sequence of 93 numbers in the interval 41–50: which we assume to be evenly spaced.

This number will be

$$\text{median} = 40.5 + \frac{68}{93} \times 10 \quad \text{(since the class width is } 50.5 - 40.5 = 10)$$

$$\simeq 48.$$

An alternative approach is to draw a cumulative frequency polygon obtained by plotting the cumulative frequencies against the upper class boundaries and joining these points by line segments. See Fig. 2.3.

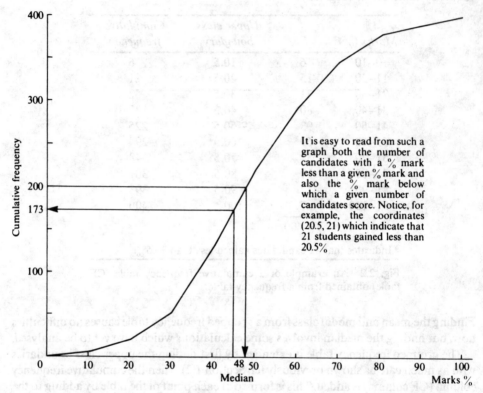

It is easy to read from such a graph both the number of candidates with a % mark less than a given % mark and also the % mark below which a given number of candidates score. Notice, for example, the coordinates (20.5, 21) which indicate that 21 students gained less than 20.5%

Fig. 2.3 A cumulative frequency polygon to show the marks gained by 400 students.

The first point to notice is the ease with which it is possible to read directly from the CF polygon an estimate of the median – a mark which 50% of the pupils fail to gain.

The number of students who pass the exam can be found just as easily. If, for example, the pass mark was 45% then reading from the graph we see that (approximately) 173 students fail and hence 227 pass. (Taking the reading from 45.5% would make little difference.)

NOTE: Joining the points in the CF polygon by line segments reflects our assumption that the data is evenly spaced in each class interval. An alternative approach is to join up the points with a smooth curve to form the *cumulative frequency curve* which can give slightly more accurate results (cf. Section 19.3).

Exercise 2.4

The CF curves should be kept, they are referred to in Exercise 4.3.

1. The table below shows the weekly wage of a group of workers.

Wage (£)	Number of workers
80 —	6
90 —	15
100 —	37
110 —	28
120–140	14

By first drawing a cumulative frequency polygon determine:
(a) the median wage;
(b) the percentage of workers who earn less than £95 per week.

2. Draw a CF polygon of the data representing the marks of 50 students taking a Mathematics examination given at the beginning of Section 2.5.
Find (a) the number of students who pass if the pass mark is 45%
(b) the pass mark if (approximately) 45% of the students pass.

3. The following grouped frequency distribution represents the marks obtained by a number of students in an examination.

Mid-interval value	Number of students
20	20
40	37
60	31
80	12

Estimate the number of students who gained more than 45 marks. Find also the limits, symmetrical about the median, between which 50% of the marks lie.

4. One hundred eggs were weighed and the results recorded as follows.

Weight (g)	less than 60	60—	65—	70—	75—	80–85
Frequency	9	18	26	30	15	2

Plot the CF polygon and hence estimate the median weight. If I buy 40 of the eggs and decide to choose the largest available, find the median weight of the eggs I purchase and the median weight of the rest.

5. Use the age distribution given in Exercise 1.1, question 12 to estimate the median age of the population of England, Wales and Scotland. Estimate the percentage of the population over 60 years of age.

Chapter 3   Probability

3.1 Introduction

In our everyday life we frequently use words such as 'certain', 'impossible' and 'probable' to describe our impression of whether a particular event will or will not occur: I am *certain* that I am alive as I write this book and will *probably* be so as you read it. Experience tells us that events have a particular chance or probability of occurring and mathematicians prefer to use a number to indicate the likelihood of, or belief in, the occurrence of a particular event.

Suppose a die is tossed a large number of times. Experience tell us that we will obtain a '2' (say) about one-sixth of the time. Alternatively, we say that the probability of obtaining a score of '2' is 1/6. Similarly the probability of obtaining a '1 or a 2' is 2/6 since our experience suggests that this event occurs twice as often.

Alternatively, in some situations, we can use symmetry or some other method to assign probabilities. In the case of a die we might argue that no one number will necessarily occur more frequently than any other when the die is tossed repeatedly. We would then assign the probability of 1/6 to each of the six possible scores. In such a case we have assumed that we are considering a 'fair' or 'unbiased' die. Naturally, if we subsequently tossed a particular die many times and found that a particular score occurred significantly more often than expected we would become suspicious. In this case we would inspect the die to check whether it was biased in some way.

We have then two methods of assigning probabilities to events. The first is to use experience, by repeating an experiment, under exactly similar conditions many times. The second is to produce a 'model' of the situation which can be used to predict what would happen if the experiment were performed. If our model proves to be acceptably accurate then we are pleased, if not then we inspect the model to check whether we can improve it.

The use of experience

Suppose we were given a die and asked to establish the probability that a '2' would be obtained on any given throw. One method then would be to toss the die a number of times and establish the value of the fraction:

$$\text{relative frequency of obtaining a } 2 = \frac{\text{number of times a '2' occurred}}{\text{number of times the die was tossed}}.$$

This method cannot be exact since different people tossing the same die could end up with different results (and they probably would!). We believe that the more often we toss the die, the more accurate our results are. In other words we obtain a better approximation to the actual probability the more often we toss the die.

A die was tossed 1000 times and the results are recorded in the following table (Fig. 3.1).

Score	1	2	3	4	5	6
Frequency	173	168	167	161	172	159
Relative frequency	0.173	0.168	0.167	0.161	0.172	0.159

Fig. 3.1

On the basis of this evidence we would argue that the probability of obtaining a '2' was approximately 0.168, say 0.17 to 2DP.

A coin was tossed 5000 times and at various stages the number of 'heads' obtained was recorded with the following results (Fig. 3.2).

Number of times coin tossed	10	20	50	100	500	1000	5000
Number of heads obtained	3	8	26	48	254	493	2471
Relative frequency	0.300	0.400	0.520	0.480	0.508	0.493	0.494

Fig. 3.2

Note that in this case the relative frequencies appear to be tending to a limit as the number of times the die is tossed is increased.

We now have a method of estimating the probability of an event.

Statement 3.1

> If an experiment is repeated n times under exactly similar conditions and an event occurs r times then the relative frequency r/n is an estimate of the probability of this event. The accuracy of this estimate increases as n increases.

The theoretical approach

Suppose we can make the assumption that a particular die is fair, which means that each number has an equal chance of occurring. In this case the probability that we obtain a '2' is 1/6.

Again, suppose a pack of cards is shuffled and the top card is turned over. We find the probability that the card is an ace by arguing that since four of the fifty-two cards in the pack are aces this probability will be 4/52.

More generally if we are interested in the probability of some particular event in a situation where there are a fixed number of possible outcomes each of which is equally likely to occur; then the probability that the event occurs equals

$$\frac{\text{number of outcomes which make the event occur}}{\text{number of possible outcomes}}$$

The phrase 'the probability that the event occurs' is often reduced to P(event). Thus $P(2)$ means the probability of obtaining a '2' (in the context of tossing a die) and P(Ace) means the probability of obtaining an Ace (in the context of choosing a card).

Example 1

Three coins are tossed and the number of 'heads' is noted. Find the probability that on any given throw two 'heads' are obtained.

METHOD 1: Toss the three coins a large number of times and calculate the estimated probability.

Exercise: Do this experiment with three coins of your own choice.

METHOD 2: List the possible outcomes of the experiment.

Outcomes	Number of 'heads'	Probability
HHH	3	1/8
HHT	2	1/8
HTH	2	1/8
THH	2	1/8
TTH	1	1/8
THT	1	1/8
HTT	1	1/8
TTT	0	1/8

Make the assumption that each coin is unbiased. (Meaning each coin is fair, that is to say, each coin has an equal chance of showing a 'head' (H) as a 'tail' (T).) Then each outcome has an equal chance of occurring. So

$$P(\text{just two 'heads'}) = 3/8.$$

Exercise 3.1 (for discussion)

1. If the probability of an event occurring is 0.14 what is the probability that it does not occur?

2. What is the probability of an event which
    (i) is certain,
    (ii) is impossible,
    (iii) has a fifty-fifty chance of occurring,
    (iv) has an evens chance of occurring?

3. An experiment can have four possible outcomes A, B, C, or D. If the probability of the first three are $P(A) = 0.12$ $P(B) = 0.16$ $P(C) = 0.35$. What is
    (i) $P(D)$,
    (ii) $P(A$ or $B)$,
    (iii) $P(\text{not } A)$?

4. It is asserted that when a drawing pin is tossed onto a smooth solid surface it lands in one of the two positions as shown here.

          A   *B*

(i) Estimate P(lands as in A) by experiment.

(ii) A young student suggests that P(lands as in A) $= 1/2$ since there are two possible outcomes and just one of these gives rise to the event required. His answer may be correct, but why is his argument fallacious?

5. In a single throw of a fair die calculate
   (i) P(an odd number occurs),
   (ii) P(a number less than 3 is obtained),
   (iii) P(a number from 1 to 6 inclusive is obtained),
   (iv) P(7 is obtained).

6. Find the probability of
   (i) failing to throw at least one 'head' with three unbiased coins,
   (ii) a man, who holds two tickets, winning a raffle for which 300 tickets have been sold.

7. A card is drawn from a well-shuffled pack of 52 cards. Find the probability that the card is
   (i) a red card, (ii) a club, (iii) a five, (iv) an ace of spades, (v) not a king, (vi) a black queen, (vii) either a king or a queen, (viii) neither a king nor a queen.

8. A die with its faces numbered 1 to 6 is biased so that the probabilities of throwing 1, 2, 3, 4, 5 are 1/12, 1/8, 1/8, 1/6, 1/4 respectively.
   (i) What is the probability of throwing a 6?
   (ii) Are you more likely to obtain an even score or an odd score?

9. A set A is defined as $A = \{-2, -1, 0, 1, 2\}$. If a number is chosen at random from A, what is the probability that it is a solution to the following?
   (i) $x^2 - 2x + 1 = 0$
   (ii) $x^2 - x - 2 = 0$

3.2 Sample spaces

The *outcome* of an experiment is, the result which actually occurs when the experiment is performed. The *sample space* is the collection of all the possible outcomes of an experiment. Naturally, one and only one of these outcomes will occur.

We can attach probabilities to each of the possible outcomes, using our previous experience, symmetry or even our own personal conviction. However, it is required that:
   (i) each probability lies between zero and one
      (0 represents the probability of an impossible outcome and 1 represents the probability of an outcome which is certain to occur);
   (ii) the sum of these probabilities is one
      (to indicate that it is certain that one of these outcomes will occur when the experiment is performed).

The choice of sample space depends not so much on the experiment as on the interests of the investigator.

Suppose two unbiased coins are tossed. A possible sample space is:

HH	HT
TH	TT

Here HT represents the outcome where a 'head' was obtained with one coin and a 'tail' was obtained with the other. In view of the fact that the coins are unbiased each of the outcomes is equally likely so that we can assign the probability 1/4 to each outcome.

However, if we were interested in the number of 'heads' obtained we might prefer to use

0,	1,	2

as our sample space. In this case the outcomes are not equally likely!

Experiment: Toss a pair of coins 100 times (more if you have the patience! A class might combine its results). Determine the relative frequencies of obtaining zero, one or two heads and hence estimate the appropriate probabilities for these outcomes.

Example 2
Two unbiased dice are tossed and the numbers noted.

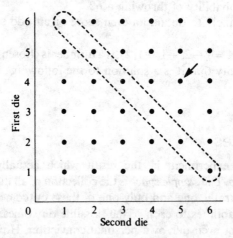

Since the dice are unbiased each outcome represented by the dots has a 1/36 chance of occurring. The arrowed dot corresponds to obtaining 4 with the first die and 5 with the second. Note that the outcome consisting of a 5 with the first die and a 4 with the second is a different outcome and has a separate dot.

Now, for example,

$$P(\text{total score of 7}) = \frac{\text{number of outcomes which result in a total score of 7}}{\text{number of possible outcomes}}$$

$$= \frac{6}{36}$$

$$= \frac{1}{6}$$

Example 3
A card is chosen from a well-shuffled pack. Find (i) $P(\text{Ace})$ (ii) $P(\text{queen or Heart})$.

(i) A suitable sample space is

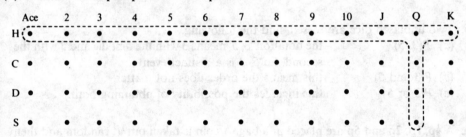

So $P(\text{Ace}) = 1/13$

(ii) A suitable sample space for this question is

So $P(\text{queen or heart}) = 16/52$

$$= 4/13$$

Note that the second sample space is suitable for answering part (i) but the first is not suitable for answering part (ii).

With a little practice it will become unnecessary to write out a sample space, but one should always be able to produce one, in principle, if required.

Exercise 3.2

1. Two unbiased dice are tossed and the scores x and y of each are noted. Find
   (a) $P(x < 3 \text{ and } y > 2)$,
   (b) $P(x = y)$,
   (c) $P(x + y = 5)$,
   (d) $P(x \cdot y = 36)$,
   (e) $P(x - y = 1)$.

2. A card is chosen at random from a pack of cards. Find
   (a) $P(3, 5)$
   (b) P(diamond or a 2),
   (c) $P(3$ or $4)$,
   (d) P(black and a king).

3. An unbiased coin and a fair die are tossed together. What is the probability of obtaining a 'head' and a 6?

4. Ten discs each with a different one of the digits $0, 1, \ldots, 9$ printed on them are placed in a bag. Two discs are taken out of the bag one at a time at random to form a two digit number, where 04 counts as 4 etc.
   (i) Assuming the first disc is replaced before the second is chosen, find the probability that
      (a) the number is even,
      (b) the number is less than 30,
      (c) the number is 67,
      (d) the two digits forming the number are equal.
   (ii) If the first disc is not replaced before the second is chosen answer the above questions.

5. Two unbiased dice are tossed, find the following.
   (a) $P(3, 5)$ the notation '3, 5' means 3 with the first die and 5 with the second. So 5, 3 is a distinct event.
   (b) $P(3$ and $5)$ this means the order does not matter.
   (c) $P(3$ or $5)$ this includes the possibility of obtaining both.

6. A $\frac{1}{2}$p, 1p, 2p and 5p are placed in a bag. A coin is taken out at random and then replaced. A second coin is taken at random and then replaced. Find the following:
   (a) $P(\frac{1}{2}p, 1p)$,
   (b) P(sum is $2\frac{1}{2}p)$,
   (c) P(at least one 5p).

7. Give an appropriate sample space for each of the following experiments.
   (a) A die is tossed repeatedly until a 'six' appears.
   (b) A football team records the results of each of two games as 'win', 'draw' or 'lose'.
   (c) An unbiased coin and a fair die are thrown together.

8. A fair die is made from a regular eight-faced solid by numbering the faces from 1 to 8. The die is thrown twice. Draw a diagram to show the sample space and calculate the probability of each of the following events.
   (a) The first score is less than 4.
   (b) The same score is obtained each throw.
   (c) The first score is greater than the second score.
   (d) The sum of the scores is 6.

9. Five cards are labelled A1, B2, C3, D3, E3 respectively. A card is selected at random and then a second card is selected again at random before the first is replaced. Show by listing that there are just 20 possible outcomes. Find the probability that
   (a) the first card chosen is A1,
   (b) the second card chosen is A1,
   (c) the card A1 is chosen,
   (d) the letter on the cards are adjacent in the alphabet,
   (e) the sum of the numbers on the cards is odd,
   (f) the sum of the numbers on the card is 5.

10. Ann and Bob each choose a number at random from the set $\{2, 3, 4, 5, 6, 7, 8\}$. Find the probability that
    (a) Ann chooses a prime number,
    (b) Ann and Bob choose the same number,
    (c) the two numbers have a sum of 9,
    (d) one chooses an even number and the other an odd number.

11. Three fair dice are thrown together.
    (a) In how many ways can they fall?
    (b) In how many ways can they fall so that the total score is 5?
    (c) What is the probability that the total score is 5?

12. Four cards are dealt from a pack of cards. If the order in which they are dealt does matter, in how many different ways can they be dealt? What is the probability that the four Aces are dealt in the order H, C, D then S?

3.3 Large sample spaces

Two particular difficulties arise when dealing with sample spaces. Firstly, the sample space can be very large and secondly the sample space can be complicated. These problems are dealt with in this and the following section.

Suppose three coins are tossed; we know that there are eight equally likely possible outcomes. If we needed to be convinced one argument is as follows.

Each outcome can be represented by three letters (each either H or T). We imagine three places to be filled according to the outcome.

outcome of first coin   outcome of second coin   outcome of third coin
         —   —   —

The first place can be filled in two ways, the second in two ways and the third in two ways; altogether there will be $2 \times 2 \times 2 = 2^3$ ways to fill the places and hence eight outcomes.

Similarly if eight coins were tossed there would be 2^8 possible outcomes. Alternatively if 10 dice were tossed there would be 6^{10} possible outcomes. However, suppose we have five cards labelled A, B, C, D and E respectively and three were chosen one at a time. Then the number of possible ways of selecting three cards is 60 rather than

$5^3 = 125$. To see that this is so imagine, as before, three spaces which are to be filled with the names of the chosen cards.

first card   second card   third card
—   —   —

The first can be filled in 5 ways (since there are five cards available to choose from). The second space can only be filled in 4 ways (since there are only four cards available to choose from). Now the third can only be filled in 3 ways as there are only three cards left. Altogether there will be $5 \times 4 \times 3 = 60$ ways to fill the places.

Again if five cards were dealt from a full pack of cards the number of possible ways of dealing the cards would be

$$52 \times 51 \times 50 \times 49 \times 48$$

Factorials

To cope with large numbers the factorial notation is introduced.

Definition 3.1

> Factorial n, written $n!$, is defined as
>
> $$n! = n \times (n-1) \times (n-2) \times \ldots \times 3 \times 2 \times 1,$$
>
> where n is a positive integer and
>
> $$0! = 1.$$

Factorials are not defined here for negative numbers.

For example

$$5! = 5 \times 4 \times 3 \times 2 \times 1$$
$$= 120$$

Useful relationships between factorials are

$$n! = n \times (n-1)! \qquad \text{for } n \geqslant 1$$
$$n! = n \times (n-1) \times (n-2)! \qquad \text{for } n \geqslant 2$$
$$n! = n \times (n-1) \times (n-2) \times (n-3)! \quad \text{for } n \geqslant 3$$

So

$$52 \times 51 \times 50 \times 49 \times 48 = \frac{52!}{47!}$$

and

$$\frac{10!}{7!} = 10 \times 9 \times 8$$

$$= 720.$$

Example 3.3

Calculate   (a) 6!   (b) $\dfrac{7!}{3!}$ (c) $\dfrac{5! \times 4!}{3! \times 6!}$

(a) $6! = 6 \times 5!$

$\qquad = 6 \times 120$

$\qquad = 720$

(b) $\dfrac{7!}{3!} = \dfrac{7 \times 6 \times 5 \times 4 \times 3!}{3!}$

$\qquad\qquad = 7 \times 6 \times 5 \times 4$

$\qquad\qquad = 840$

(c) $\dfrac{5! \times 4!}{3! \times 6!} = \dfrac{5! \times 4 \times 3!}{3! \times 6 \times 5!}$

$\qquad\qquad\quad = \dfrac{4}{6}$

$\qquad\qquad\quad = \dfrac{2}{3}$

Example 3.4

Write in factorial notation

(a) $40 \times 39 \times 39$

(b) $\dfrac{12 \times 11 \times 10}{6 \times 5 \times 4}$

(a) $40 \times 39 \times 39 = \dfrac{40!}{38!} \times 39$

$\qquad\qquad\qquad\quad = \dfrac{40! \times 39!}{(38!)^2}$

(b) $\dfrac{12 \times 11 \times 10}{6 \times 5 \times 4} = \dfrac{12! \times 3!}{9! \times 6!}$

Statement 3.2

Generally if one event can occur in e_1 ways, a second in e_2 ways,..., an nth event in e_n ways, then the whole set can occur in $e_1 \times e_2 \times \ldots \times e_n$ ways.

Example 3.5
A bag contains five balls, each a different one of the colours red, white, black, green and yellow. Balls are taken out one at a time, the colour noted, and then *replaced*. How many distinct outcomes are there if we consider (a) the first two taken, (b) the first five taken?

(a) There are five ways of selecting the first ball and five ways of selecting the second so altogether there are $5 \times 5 = 5^2 = 25$ distinct outcomes.
(b) In this case there are 5^5 distinct outcomes.

If we suppose that the balls are *not replaced* then:
(a) There are still five distinct ways of selecting the first ball, but only four ways of selecting the second since the first ball is not replaced. So there will be $5 \times 4 = 20$ distinct outcomes.
(b) In this case we have $5 \times 4 \times 3 \times 2 \times 1 = 5!$ distinct outcomes.

The reader should bear in mind whether he is 'sampling' with or without replacement.

Example 3.6
(a) Four dice are tossed. Find the probability that four sixes occur.
(b) A child arranges four cards, each displaying one of the letters G. I. R. L., in a row. Find the probability that the cards are arranged in the order GIRL.

(a) The number of possible outcomes is 6^4, just one of which gives rise to four sixes, so

$$P(6, 6, 6, 6,) = \frac{1}{6^4} \text{ assuming the dice are unbiased.}$$

(b) The number of possible outcomes is 4! just one of which gives rise to the word GIRL.

$$P(G,I,R,L) = \frac{1}{4!} \text{ assuming the child orders the cards at random.}$$

Exercise 3.3

1. Write down the following in factorial notation.

(a) $6 \times 5 \times 4$ (b) $n \times (n-1)$ (c) $\dfrac{9 \times 8 \times 7}{3 \times 2 \times 1}$

(d) $n(n-1)(n-2)(n-3)$ (e) $n(n-1)(n-2) \times \ldots \times (n-r+1)$

(f) $\dfrac{n(n-1)(n-2) \times \ldots \times (n-r+1)}{r(r-1)(r-2) \times \ldots \times 2 \times 1}$

2. Evaluate the following.   (a) $6!$ (b) $\dfrac{7!}{5!}$ (c) $\dfrac{19!}{16!}$ (d) $\dfrac{8!}{5! \, 3!}$

3. Simplify.

(a) $\dfrac{(n+1)!}{(n-1)!}$ (b) $\dfrac{(n+2)!}{n!}$ (c) $\dfrac{(n-1)!}{(n-2)!}$

4. Eight unbiased coins are tossed, find the probability that
   (a) eight heads are obtained,
   (b) just seven heads are obtained.

5. A combination lock has four rings each printed with the 10 digits 0, 1, ..., 9. Find the probability of finding the correct combination if the digits are chosen at random.

6. Given the (unlikely) assumption that the order in which eight horses complete a race, is random, find the probability that a gambler can correctly
   (a) place the first three horses,
   (b) guess the first three home.

7. Four boys and three girls sit at random on a bench. Find the probability that
   (a) the boys alternate with the girls,
   (b) the boys sit on the left of the girls.

8. Find the probability that of four people chosen at random
   (a) all four have their birthdays in December,
   (b) all four have their birthdays in the same month,
   (c) no two of the people have their birthdays during the same month.
   You may assume that people are equally likely to have their birthdays in any month.

9. Discuss whether the assumption above is reasonable.

10. The letters in the following words are rearranged at random. Find the probability that when rearranged again at random they form the original word.
    (a) SAMPLE,   (b) TREE.

3.4 Permutations and combinations

The solution to problem 6(a), Exercise 3.3. can be generalised.

The number of ways of arranging or *permuting* n distinct objects is n! Thus there are six ways (3!) of permuting the letters A, B and C as follows

$$\text{ABC, BCA, CAB, ACB, BAC, CBA.}$$

The number of ways of permuting (arranging) just r objects from n distinct objects can be computed as follows.

Imagine the r spaces to be filled

$$\underline{\quad}\ \underline{\quad}\ \underline{\quad}\ \underline{\quad}\ \cdots\ \underline{\quad}$$

There are n ways to fill the first space and $(n - 1)$ ways to fill the second. Then there will be $(n - 2)$ ways to fill the third and so on until we find there are $(n - r + 1)$ ways to fill the rth space. Altogether there are just

$$n \times (n - 1) \times (n - 2) \times \dots \times (n - r + 1) \text{ ways}$$

to permute r objects from n. But this is precisely

$$\frac{n!}{(n - r)!}.$$

For convenience we make the following definition.

Definition 3.2

> The number of ways of permuting r objects from n distinct objects written as ^nP_r is
>
> $$^nP_r = \frac{n!}{(n - r)!}$$

Problem 6(a), Exercise 3.3 can now be solved as follows.

The number of ways of permuting 3 horses from 8 is

$$^8P_3 = \frac{8!}{5!} = 8 \times 7 \times 6 = 336$$

Just one of these possible outcomes can be correct and so the probability is $1/336$.

Suppose now that we are simply interested in the number of ways of choosing (selecting or combining) r objects from n distinct objects; in this case the order of selection is immaterial. This was the situation in problem 6(b), Exercise 3.3.

Consider the number of ways of permuting three letters from A, B, C and D. Since

$$^4P_3 = \frac{4!}{(4 - 3)!} = 24,$$

there are 24 distinct permutations. These are listed below

ABC	ABD	ACD	BCD
BCA	BDA	CDA	CDB
CAB	DAB	DAC	DBC
CBA	DBA	DCA	DCB
BAC	ADB	ADC	BDC
ACB	BAD	CAD	CBD

An inspection of the display above shows that there are just *four* different ways of choosing three objects from four. The first entry ABC, for example, has six permutations ($^3P_3 = 6$), similarly for each of the other entries in the top row. Once the *combination* or choice of the three letters A, B, C, has been made, further permutations will not alter the combination of letters.

When we permuted three objects from four in an attempt to find the number of ways of combining or choosing three objects from four we overcounted by a factor of 6 ($^3P_3$). So the required number of combinations is

$$4 = \frac{24}{6} = \frac{^4P_3}{3!}$$

Similarly the number of combinations of four objects from six is

$$\frac{^6P_4}{4!} = \frac{6!}{(6-4)!\,4!} = \frac{6.5}{2!} = 15$$

In general

Definition 3.3

The number of distinct ways of choosing a set or combination of r objects from n, written as ^nC_r, is

$$^nC_r = \frac{^nP_r}{r!} = \frac{n!}{(n-r)!\,r!}$$

Some authors write ^nC_r as nCr or $\binom{n}{r}$.

Example 3.5
Evaluate   (a) $^{16}C_{14}$ (b) $^9C_4$.

(a) $^{16}C_{14} = \frac{16!}{2!\,14!} = \frac{16 \times 15}{2} = 120$

(b) $^9C_4 = \frac{9!}{5!\,4!} = \frac{9 \times 8 \times 7 \times 6}{4 \times 3 \times 2 \times 1} = 126$

Example 3.6
Show that $^nC_r = {}^nC_{n-r}$

$$^nC_{n-r} = \frac{n!}{(n-(n-r))!\,(n-r)!}$$

$$= \frac{n!}{r!\,(n-r)!}$$

$$= \frac{n!}{(n-r)!\,r!}$$

This is a reasonable result since every choice of r objects from n results in leaving $n-r$ objects behind. This is equivalent to choosing $n-r$ objects to leave behind. Thus the number of ways of choosing r objects from n is the same as the number of ways of choosing $n-r$ objects from n.

    A related problem arose in question 10, Exercise 3.3. The number of ways of permuting four distinct letters is $4 \times 3 \times 2 \times 1 = 4!$ ($^4P_4$) but if two or more letters are the same and hence indistinguishable, there must be fewer permutations.

Consider the word 'BEER'. If we distinguish between the two E's by calling the first E_1 and the second E_2 then there are twenty-four (4!) possible 'words'. However, since there is no distinction we can rearrange the two E's in each word without altering it, and so we have overcounted the number of words by a factor of 2 (the number of ways of permuting the two E's, $^2P_2$) so the actual number is just 12.

BEER	EBER	EEBR	REBE	EBRE	BREE
REEB	EREB	EERB	BERE	ERBE	RBEE

Statement 3.3

More generally, if there are n objects with r_1 alike and another r_2 alike and so on, then there are

$$\frac{n!}{r_1! \times r_2! \times \ldots}$$

ways of permuting these objects

Example 3.7

Find the number of distinct permutations of the letters in the word 'PROBABILITY'.

There are 11 letters in the word, but 'B' appears twice and 'I' appears twice, so the number of distinct permutations is $\dfrac{11!}{2!\,2!}$.

Example 3.8

An unbiased coin is tossed six times. Find

(a) the number of possible outcomes,

(b) the probability that first three 'tails' are obtained and then three 'heads',

(c) the probability that just three 'heads' are obtained.

(a) For each toss there are two possible outcomes so altogether there are $2^6 = 64$ outcomes.

(b) Just one of the possible outcomes gives rise to TTTHHH as required so the probability is $\dfrac{1}{64}$.

(c) To find the number of ways of obtaining just three heads we imagine choosing three of the six tosses to be 'heads' we can do this in

$$^6C_3 = \frac{6!}{3!\,3!} \text{ ways}$$

$$= \frac{6 \times 5 \times 4}{3 \times 2 \times 1}$$

$$= 20$$

Thus, the probability of obtaining just three heads is

$$\frac{20}{64} = \frac{5}{16}$$

Example 3.9
A committee of 5 is to be chosen from 6 men and 4 women. Find the probability that exactly 4 men are chosen; assuming that each person is equally likely to be selected.

The number of ways of choosing 5 people from ten is

$$^{10}C_5 = \frac{10 \times 9 \times 8 \times 7 \times 6}{5 \times 4 \times 3 \times 2 \times 1} = 252$$

If exactly 4 men are chosen then there must be exactly 1 woman as well. The number of ways of doing this is

$$^6C_4 \times {}^4C_1 = \frac{6!}{1!\,4!} \times \frac{4!}{3!\,1!}$$

$$= 15 \times 4$$

$$= 60.$$

So the required probability is

$$\frac{60}{252} = \frac{5}{21}.$$

Exercise 3.4
1. Evaluate

    (a) $^5P_3$ (b) $^6C_4$ (c) $^6P_0$

    (d) $^7C_0$ (e) $^7C_1$ (f) $^8C_2$

    (g) $^8C_6$ (h) $^{19}C_{16}$ (i) $^nC_1$

    (j) ^nP_n (k) $^nP_2$ (l) $^{12}C_1$

2. In how many ways is it possible to select four objects from seven if
    (a) the order matters,
    (b) the order is irrelevant.

3. Show by calculation that $^8C_5 = {}^8C_3$.
    Explain why the number of ways of choosing five objects from eight is the same as the number of ways of choosing three objects from eight.
    Prove that $^mC_p = {}^mC_{m-p}$.

4. Write out in ^nC_r notation

    (a) $\dfrac{(n+m)!}{m!\,n!}$,   (b) $\dfrac{(n+1)!}{(n-r+2)!\,(r-1)!}$.

5. How many permutations of the letters in the word 'STATISTICS' are possible?

6. Twelve people each spin a coin. Find the number of ways exactly 5 heads may be obtained. Assuming the coins are unbiased find the probability of obtaining exactly five heads.

7. In how many ways can a committee of 4 men and 3 women be formed from 7 men and 8 women?
   If the committee was chosen at random find the probability of including both Jack and his wife.

8. If $3 \times {}^{2n}C_3 = 44 \, {}^{n}C_2$ find the value of n.

9. A bridge hand consists of 13 cards, find the probability that a player obtains all four aces.

10. Repeat questions 6, 8 and 10 in Exercise 3.3.

11. In how many ways can three distinct balls be placed in six containers placed in a row
    (a) so that no container receives more than one ball,
    (b) without restrictions.

12. A boy keeps three pairs of black and four pairs of brown socks in a drawer.
    (a) If he takes two socks at random, what is the probability that they are the same colour?
    (b) If he takes three at random what is the probability that at least two are of the same colour?

3.5 Tree methods

An extremely useful way of dealing with complicated sample spaces is now introduced. Only a partial justification of the method is given at this stage, but it will be justified in Chapter 5.
   Suppose a bag contains three red balls and four white balls. A ball is taken out at random, its colour noted and then replaced. The experiment is then repeated. There are four possible situations or events which can arise.

$$\text{RR} \qquad \text{WR}$$

$$\text{RW} \qquad \text{WW}$$

where, for example, RW represents the event of obtaining a red ball first and then a white ball second. We shall be interested in finding the following:
(a) $P(R, R)$,
(b) $P(W, R)$,
(c) P(just one of the two balls chosen is red),
(d) P(the second ball taken is red).

We might be tempted to use the following diagram to represent a suitable sample space

First	R	•	•
choice	W	•	•
		R	W

Second choice

We soon realise that we cannot readily assign probabilities because the dots do not represent equiprobable outcomes since there are more white balls than red.

The following is a suitable sample space (Fig. 3.3).

	W	•	•	•	•	•	•	•
	W	•	•	•	•	•	•	•
First	W	•	•	•	•	•	•	•
Choice	W	•	•	•	•	•	•	•
	R	•	•	•	•	•	•	•
	R	•	•	•	•	•	•	•
	R	•	•	•	•	•	•	•
		R	R	R	W	W	W	W

Second choice

Fig. 3.3.

Clearly

(a) $P(R, R) = \dfrac{9}{49}$

(b) $P(W, R) = \dfrac{12}{49}$

(c) $P(\text{just one red}) = \dfrac{24}{49}$

(d) $P(\text{second red}) = \dfrac{21}{49}$

$$= \dfrac{3}{7}$$

Consider the following diagram called a TREE DIAGRAM

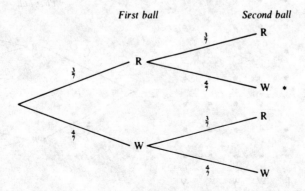

Fig. 3.4

Such a tree diagram is useful when events can be subdivided into an ordered sequence of sub events. Each 'branch' of the tree then represents one of the possible events. The 'branch' marked * in Fig. 3.4 represents the event of obtaining a red ball first and a white ball second.

The probabilities entered on the lines represent the probabilities of the sub events. Thus for the 'branch' marked * the probability that the first ball is red is 3/7 and the probability that the second ball is White is 4/7.

Statement 3.4

> The probability of any event represented by a single branch of the tree is the product of the probabilities along this branch.

Statement 3.5

> The probability of any event which involves two or more 'branches' is the sum of the probabilities of each 'branch'.

A study of Fig. 3.3 should ensure that the reader finds these statements reasonable. Proofs are given in Chapter 5.

Using this 'tree method' we obtain

(a) $P(R, R) = \dfrac{3}{7} \times \dfrac{3}{7} = \dfrac{9}{49}$ ⎫ considering just one 'branch' in each case

(b) $P(W, R) = \dfrac{4}{7} \times \dfrac{3}{7} = \dfrac{12}{49}$ ⎭

(c) $P(\text{just one } R) = \dfrac{3}{7} \times \dfrac{4}{7} + \dfrac{4}{7} \times \dfrac{3}{7} = \dfrac{24}{49}$ ⎫ considering two of the 'branches' in each case

(d) $P(\text{second is } R) = \dfrac{3}{7} \times \dfrac{3}{7} + \dfrac{4}{7} \times \dfrac{3}{7} = \dfrac{21}{49}$ ⎭

which agrees with the previous method.

If the balls were *not replaced* after each choice the appropriate outcome space and tree diagram are as shown in Fig. 3.5.

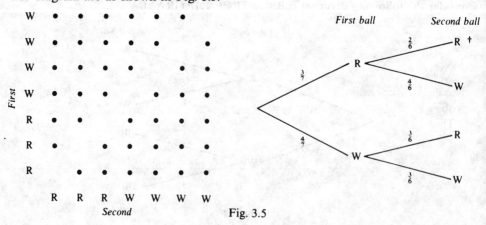

Fig. 3.5

The difference in this case is due to the fact that the same ball cannot be chosen twice.

To calculate the probabilities along the 'branches' in the tree diagram in Fig. 3.5 account must be taken of the fact that the balls are not replaced. Consider the event RR represented by the branch marked †. The probability that the first ball is red is 3/7 since there are three red balls among the seven in the bag; but now, for the second choice, there are only six balls and just two of them are red, (if the first ball taken was red) so in this case the probability that the second ball is red is 2/6. Similarly for the other probabilities. Notice again that Statements 3.4 and 3.5 still hold.

Example 3.10

A die is tossed three times, find the probability that the first score is even, the second is odd and the third is either a one or a two. Find also the probability that just one six is obtained.

A suitable 'tree' to consider for the first part would be as follows.

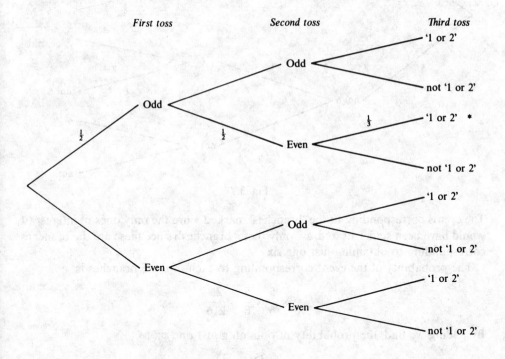

Fig. 3.6

The event of interest corresponds to the single branch marked * and so only the probabilities along this branch have been entered.

Thus

$$P(\text{even, odd, 1 or 2}) = \frac{1}{2} \times \frac{1}{2} \times \frac{1}{3}$$

$$= \frac{1}{12}.$$

This tree would not have been suitable to find the probability that all three scores were even, though the reader could readily provide a suitable tree. With practice it is not even necessary to draw the tree to find the required probabilities.

For the second part an appropriate tree is as shown in Fig. 3.7.

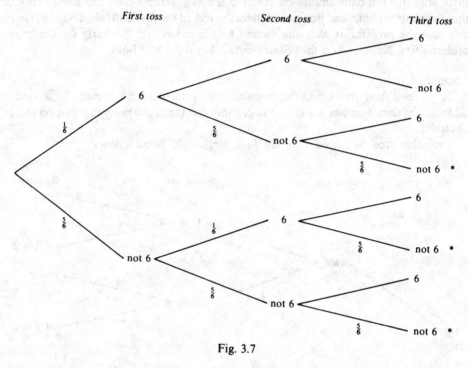

Fig. 3.7

The events corresponding to the 'branches' marked * are the only ones of interest (it would have been sufficient to draw only these 'branches') since these are the branches corresponding to obtaining just one six.

The probability of the event corresponding to each of these branches is

$$\frac{1}{6} \times \frac{5}{6} \times \frac{5}{6} = \frac{25}{216}$$

By adding we find, the probability of obtaining just one six as

$$3 \times \frac{25}{216} = \frac{75}{216}$$

$$= \frac{25}{72}$$

$$\simeq 0.35$$

Note the use of a decimal to represent a probability. The main advantage of this is that one can more readily interpret a probability when it is written as a decimal rather than a complicated fraction.

Example 3.11
In a garden there are ten red flowers and six blue ones. If two flowers are selected at random find the probability that they are
(a) both blue,
(b) different in colour.

In order to use a 'tree' method in this case we need to impose an order into the selection of the flowers and suppose that one flower is picked first, then the second is selected. A suitable tree is now found in Fig. 3.8.

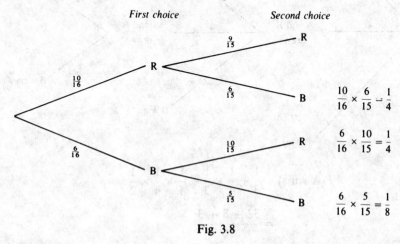

Fig. 3.8

Care needs to be exercised as usual when filling in the probabilities. By way of an example, consider the bottom 'branch' corresponding to BB. Here the probability that the first choice is a blue flower is 6/16 since six of the sixteen flowers are blue. But the probability that the second flower is blue is now reduced since there are only fifteen flowers to choose from and only five of them are blue (assuming the first chosen is blue).

(a) $P(\text{BB}) = \dfrac{1}{8}$

(b) $P(\text{different colour}) = \dfrac{10}{16} \times \dfrac{6}{15} + \dfrac{6}{16} \times \dfrac{10}{15}$

$$= \frac{1}{4} + \frac{1}{4}$$

$$= \frac{1}{2}$$

Example 3.12
Two boys decide to enter a chess competition. The competition is won by the first to score 1 point or more, and is decided by the toss of a coin if they reach 1 point simultaneously. Each scores 1 point for winning a particular game, 1/2 for a draw and 0 if the game is lost.

If the first boy has a probability of 1/3 of winning a game and a probability of 1/4 of gaining a draw, find the probability that the first boy wins the competition.

Suppose the first boy is called Anatoly and the second Boris,

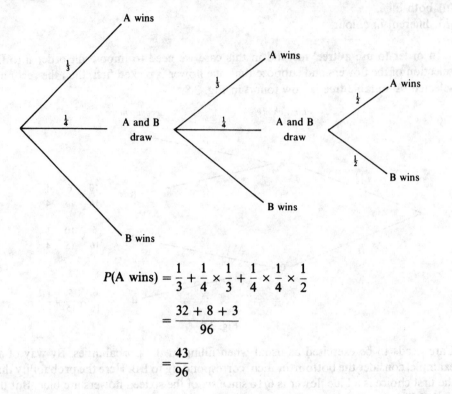

$$P(\text{A wins}) = \frac{1}{3} + \frac{1}{4} \times \frac{1}{3} + \frac{1}{4} \times \frac{1}{4} \times \frac{1}{2}$$

$$= \frac{32 + 8 + 3}{96}$$

$$= \frac{43}{96}$$

Exercise 3.5

1. Repeat the example introduced at the beginning of Section 3.5 assuming that the balls are not replaced.

2. Two dice are loaded so that $P(6) = 1/2$ and the probabilities that each of the other numbers occurs are equal. If both dice are tossed find
   (a) $P(6, 4)$, (b) $P(6 \text{ and } 4)$, (c) $P(6,6)$, (d) $P(\text{total score is more than 9})$.

3. For this question find the answers by (i) permutation and combination methods; (ii) tree methods.
   (a) Three cards are drawn at random from a full pack without replacement. Find the probability that they are J, 4 and K in that order.
   (b) If two cards are chosen without replacement (with Ace low) what is the probability that they are consecutive cards of the same suit?

4. A bag contains a dozen apples, of which three are bad. If two apples are drawn at random, find the probability that
   (a) both are good,

(b) both are bad,

(c) one is good, the other bad.

5. A packet of sweets contains 5 toffees, 4 mints and 3 fruit gums. Find the probability that

(a) two sweets chosen at random will be toffees,

(b) of three sweets chosen at random the first two will be the same and the third a toffee,

(c) the three gums will be chosen first.

6. From an ordinary pack of cards the picture cards (not aces) are selected and the rest discarded. From these twelve cards various cards are chosen at random.

(i) If two cards are selected, find the probability that they are both kings.

(ii) If the first card was a King, then find the probability that the second is

(a) another King,

(b) a Queen.

7. Four red discs, numbered 1 to 4 respectively and five green discs, numbered 1 to 5 respectively are placed together in a bag. Various discs were drawn at random.

(i) If just one disc is drawn what is the probability

(a) that it is red,   (b) that it has a 3 on it?

(ii) If two discs are drawn together find the probability that

(a) they are both the same colour,

(b) the sum of the numbers they bear is 7.

8. John and Brian each have three coins. John tosses each of his coins and gives to Brian those which fall heads. Brian now tosses each of the coins in his possession, giving to John those which fall heads.

Find (a) P(John has all the coins),

(b) P(Brian has all the coins),

(c) P(Both have the same number of coins).

9. A bag contains 6 red and 5 blue beads. Four are drawn out at random and not replaced. Find the probability that

(a) they are all red,

(b) there is just one red bead.

10. An unbiased die is thrown four times.

(a) Find the probability that at least one six is obtained.

(b) Answer part (a) if the die is tossed n times.

11. A bag contains m red balls and n blue balls which are taken out one by one at random. Find the following probabilities

(a) P(R)

(b) P(R, B)

(c) P(R and B)

(d) P(the second is R if the first is B).

Chapter 4   Measures of spread

4.1 Leading to the standard deviation

In a series of five short mathematics tests John and Peter score as follows:

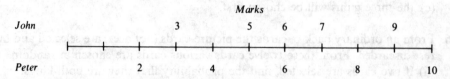

We can easily establish that each has the same mean and median, but inspection of the figures indicates that John is more consistent than the rather erratic Peter! It would be useful to have a measure of spread of the marks in each case: a number which would indicate the extent to which the sequence of numbers differs from the mean (say).

Range

This measure of spread is obtained by subtracting the smallest number from the largest number.

	Range
John	6
Peter	8

The result agrees with our view that John is more consistent and so has a smaller spread of marks.

Advantages
This measure is easy to calculate.

Disadvantages
The range depends only on two extreme values. Suppose Mary and Ann gained the following marks.

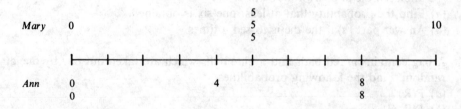

In this case we have

	Range
Mary	10
Ann	8

and this conflicts with our intuitive view of the spread.

Mean absolute deviation from the mean

Clearly it would be better to make use of more of the information. One suggestion might be to subtract the mean from each number to obtain the *deviation from the mean* and then (say) add them together.

In John and Peter's case we have

	Deviation from the mean					
John	−3	−1	0	1	3	Sum 0
Peter	−4	−2	0	2	4	Sum 0

which might be what we expected, but not what we wanted!

A better idea would be to ignore the negative signs, since in trying to measure the 'spread' we are not concerned whether the numbers are above or below the mean, just in how far they are from the mean. These are the *absolute* deviations from the mean.

	Absolute deviation from the mean					
John	3	1	0	1	2	Sum 7
Peter	4	2	0	1	4	Sum 12

So, like the range, these numbers agree with our impression of the spread. In Mary and Ann's case we have

	Absolute deviation from the mean					
Mary	5	0	0	0	5	Sum 10
Ann	4	4	0	4	4	Sum 16

which is more acceptable.

But what of Paul, who was absent for one of the tests, with the following marks?

	Marks				
Paul	2	4	5	9	Mean 5
	Absolute deviation from the mean			
Paul	3	1	0	4	Sum 8

In this case his lower sum is in part due to there being fewer marks to add.

The obvious solution is to consider not the sum of the absolute deviations, but rather the mean of the absolute deviations.

Finally we have

Mean absolute deviation from the mean	
John	1.4
Peter	2.4
Mary	2.0
Ann	3.2
Paul	2.0

Advantages

This measure seems to agree with our intuitive idea of a measure of spread and certainly involves all of the information.

Disadvantages

This measure is difficult to deal with theoretically (as will be evident in later algebraic work). Formally

Definition 4.1

$$\text{Mean absolute deviation from the mean} = \frac{\Sigma|x - \bar{x}|}{n}$$

where n is the number of item, $\bar{x}$ the mean of the numbers, Σ the sum of all the terms $|x - \bar{x}|$ and the modulus sign $|\ .\ |$ indicates the positive value. It is the modulus sign which causes the difficulties.

Variance and standard variation

The modulus sign was introduced above in order to eliminate the negative signs in the deviation from the mean. An alternative approach is to square each number and then continue as before and find the mean of these squared numbers.

So, in the case of John's marks we have

Marks	x	3	5	6	7	9	$\bar{x} = 6$
Deviations from the mean	$x - \bar{x}$	-3	-1	0	1	3	
Square of the deviation	$(x - \bar{x})^2$	9	1	0	1	9	
Mean of squared deviation from the mean	$\dfrac{\Sigma(x - \bar{x})^2}{n} = \dfrac{20}{5} = 4$						

We call this measure the *variance*. Thus

Definition 4.2

$$\text{Variance} = \frac{\Sigma(x - \bar{x})^2}{n} = \frac{\text{mean (of the) squared}}{\text{deviations from the mean}}$$

Advantages
This measure is excellent for theoretical work.

Disadvantages
The units are curious. If we were considering not marks, but lengths (say) in cm, then the mean, mode, median, range and mean absolute deviation from the mean would all be in cm; yet the variance is measured in cm^2. Because of this we make the following definition.

Definition 4.3

$$\text{Standard deviation} = \sqrt{\text{Variance}}$$

$$S_x = \sqrt{\frac{\Sigma (x - \bar{x})^2}{n}}$$

Root mean squared deviation

It is easy to see that

	Standard deviation
John	2
Peter	2.83
Mary	3.16
Ann	3.58
Paul	2

The reader may feel that he has now met so many measures of spread that he does not know which is the 'correct' one to use. In a sense it does not matter. We simply want any measure which agrees to some extent with our intuitive ideas of spread and which we can deal with mathematically. Both the variance and the standard deviation satisfy these criteria.

The actual number representing the spread does not really matter. We use it for purposes of comparison. So, for example, a measure of spread which is large compared with the mean gives us a different picture in our minds of the data than a measure which is small compared with the mean. Again, of two sets of data, if one has a larger measure of spread than the other then we know that the former is more scattered than the latter.

Exercise 4.1

1. Check the standard deviations recorded in the table above.

2. Find the mean, mean absolute deviation from the mean, the variance and the standard deviation of each of the following sets of data:
   (a) 1 2 5 6 8 14;

(b) 1   7   8   9   10   12   12   15   19   19;

(c) the first ten natural numbers (1–10 inclusive).

3. (a) Find the variance of 40, 41, 42, 43, 46.

   (b) Subtract a number M (say) of your own choice from each of the above numbers
       to obtain a new sequence. Naturally this will reduce the mean of the numbers
       by the number M, but check that it does not alter the variance. This is obviously
       a desirable property of any measure of spread.

4. The variance, $\Sigma (x - \bar{x})^2/n$, is the mean squared deviation from the mean. Now
   $\Sigma (x - M)^2/n$ is called the mean squared deviation about M. Calculate the mean
   squared deviation about 10 for the following.

$$7 \quad 9 \quad 12 \quad 16$$

   Show that in this case

$$\text{Variance} = (\text{mean squared deviation about 10}) - (10 - \bar{x})^2$$

5. In general

$$\text{Variance} = (\text{mean squared deviation about } A) - (A - \bar{x})^2$$

   Using this result show that the variance is the smallest mean squared deviation.

6. The mean squared deviation about zero is

$$\frac{\Sigma (x - 0)^2}{n} = \frac{\Sigma x^2}{n} = \overline{x^2} \text{ (the mean of the } x^2)$$

   Use the result given in question 5 to show that

$$\text{Variance} = \overline{x^2} - \bar{x}^2.$$

   Repeat question 2 using the above result to perform the calculation of the
   variance.

7. By first subtracting a suitable number and then using the result given in question 6
   above find the standard deviation of
   (a) 126   129   131   133   135;
   (b) 51   51   51   52   52   53   53   53   53   55.

8. Explain why the standard deviation of any set of numbers is never negative.

9. Let $\bar{x}$ be the mean of the sequence $x_1, x_2, x_3, \ldots, x_n$. Show that by subtracting $\bar{x}$ from
   each of the numbers the new sequence obtained will have the same variance, but
   zero mean.

10. Find the mean absolute deviations for the following

$$4 \quad 4 \quad 3 \quad 2 \quad 3 \quad 4 \quad 2 \quad 4 \quad 1$$

(a) from the mode;
(b) from the median;
(c) from the mean.

4.2 Measure of spread from frequency tables

In question 6 of Exercise 4.1 we indicated that

$$\text{Variance} = \overline{x^2} - \bar{x}^2$$

i.e. 'the mean of the squares minus the square of the means'

We use this formula to find the variance of the data given in the following frequency distribution.

x	f	fx	fx^2
1	3	3	3
2	2	4	8
3	3	9	27
4	0	0	0
5	2	10	50
	$\Sigma f = 10$	$\Sigma fx = 26$	$\Sigma fx^2 = 88$

$\text{Variance} = \overline{x^2} - \bar{x}^2$

$$= \frac{\Sigma fx^2}{\Sigma f} - \left(\frac{\Sigma fx}{\Sigma f}\right)^2$$

$$= \frac{88}{10} - \left(\frac{26}{10}\right)^2$$

$$\simeq 2.04$$

The working is simply an extension of that required for finding the mean. The additional fx^2 column is obtained by multiplying the terms in the x column by the corresponding terms in the fx column.

When data are given in a frequency distribution format the formal definition of variance is

Definition 4.4

$$\text{Variance} = \frac{\Sigma f(x - \bar{x})^2}{\Sigma f}$$

$$\text{which we claim} = \frac{\Sigma fx^2}{\Sigma f} - \left(\frac{\Sigma fx}{\Sigma f}\right)^2$$

$$= \overline{x^2} - \bar{x}^2$$

The equality can be proved as follows.

Let $\quad \Sigma f = n$

$$\frac{\Sigma f(x - \bar{x})^2}{n} = \frac{\Sigma f(x^2 - 2x\bar{x} + \bar{x}^2)}{n}$$

$$= \frac{\Sigma fx^2}{n} - \frac{\Sigma f.2x\bar{x}}{n} + \frac{\Sigma f\bar{x}^2}{n}$$

$$= \frac{\Sigma fx^2}{n} - 2\bar{x}\frac{\Sigma fx}{n} + \bar{x}^2\frac{\Sigma f}{n}$$

$$= \overline{x^2} - 2\bar{x}^2 + \bar{x}^2$$

$$= \overline{x^2} - \bar{x}^2$$

4.3 Coding

When considering the mean we introduced coding where

$$x = M + Cu$$

and proved that

$$\bar{x} = M + C\bar{u}.$$

The corresponding results for the variance and standard deviation are

Statement 4.1

$$\boxed{\begin{array}{c} \text{Var}(x) = C^2 \, \text{Var}(u) \\ S_x = C \, S_u \end{array}}$$

where Sx means the standard deviation of the x values. Similarly Su means the standard deviation of the u values.

These results can be proved as follows.

$\text{Var } x = \text{Var } (M + Cu) \qquad\qquad$ since $x = M + Cu$

$\displaystyle = \frac{\Sigma f(M + Cu - \overline{(M + Cu)})^2}{\Sigma f} \qquad$ by definition

$\displaystyle = \frac{\Sigma f(M + Cu - (M + C\bar{u}))^2}{\Sigma f} \qquad$ rewriting

$\displaystyle = \frac{\Sigma f(Cu - C\bar{u})^2}{\Sigma f}$

$\displaystyle = C^2 \frac{\Sigma f(u - \bar{u})^2}{\Sigma f} \qquad\qquad$ since C is a constant in the summation

$= C^2 \, \text{Var } u$

The second result follows immediately.

Example 4.1

Find the mean and the standard deviation of the data in the following grouped frequency distribution representing the distribution of marks in an examination.

Marks	f	Mid-interval x	$u = \dfrac{x - 45}{10}$	fu	fu^2
0 —	1	5	−4	−4	16
10 —	3	15	−3	−9	27
20 —	8	25	−2	−16	32
30 —	11	35	−1	−11 (−40)	11
40 —	17	$M \rightarrow 45$	0	0	0
50 —	36	55	1	36	36
60 —	27	65	2	54	108
70 —	16	75	3	48	144
80 —	0	85	4	0	0
90–100	1	95	5	5 (143)	25
	$\Sigma f = 120$			$\Sigma fu = 103$	$\Sigma fu^2 = 399$

Thus

$$\bar{u} = \frac{\Sigma fu}{\Sigma f} = \frac{103}{120} = 0.858$$

$$\text{Var}(u) = \overline{u^2} - \bar{u}^2 = \frac{399}{120} - \left(\frac{103}{120}\right)^2 = 2.588$$

$$S_u = \sqrt{2.588} = 1.609$$

Now since

$$x = 45 + 10u$$

$$\bar{x} = 45 + 10\bar{u}$$

$$= 53.58$$

$$\simeq 53.6$$

$$Sx = 10\,Su$$

$$= 16.09$$

$$\simeq 16$$

Example 4.2

A set of four numbers has a mean of 3 and variance of 2 and a further set of six numbers has a mean of 7 and a variance of 5. Find the mean and variance of the set of ten numbers.

Suppose the four numbers in the first set are

$$x_1, x_2, x_3, x_4$$

then

$$\text{mean} = \frac{\Sigma x}{4} = 3$$

so

$$\Sigma x = 12$$

and

$$\text{variance} = \frac{\Sigma x^2}{4} - \left(\frac{\Sigma x}{4}\right)^2 = 2$$

$$\frac{\Sigma x^2}{4} - 9 = 2$$

$$\Sigma x^2 = 44$$

Similarly if the six numbers in the second set are

$$x_5, x_6, x_7, x_8, x_9, x_{10}$$

then

$$\text{mean} = \frac{\Sigma x}{6} = 7$$

$$\Sigma x = 42$$

and

$$\text{variance} = \frac{\Sigma x^2}{6} - \left(\frac{\Sigma x}{6}\right)^2 = 5$$

$$\frac{\Sigma x^2}{6} - 49 = 5$$

$$\Sigma x^2 = 324$$

To find the mean and variance of the two sets taken together, we shall require Σx and Σx^2 where the summation is taken over all ten numbers.

Now

$$\Sigma x = 12 + 42$$

$$= 54$$

so the mean is 5.4 and

$$\Sigma x^2 = 44 + 324$$

$$= 368$$

so

$$var(x) = \frac{\Sigma x^2}{10} - \left(\frac{\Sigma x}{10}\right)^2$$

$$= \frac{368}{10} - 5.4^2$$

$$= 7.64$$

Example 4.3
A batch of washers has the following distribution of diameters.

Diameter in cm	1.250—	1.252—	1.254—	1.256—	1.258—	1.260—	1.262—	1.266—1.266-1.268
Number of washers	2	3	5	14	23	33	13	6	1

Find the mean and standard deviation of the above frequency distribution.

Class interval	f	x	u	fu	fu²
1.250 —	2	1.251	-4	-8	32
1.252 —	3	1.253	-3	-9	27
1.254 —	5	1.255	-2	-10	20
1.256 —	14	1.257	-1	$-14\,(-41)$	14
1.258 —	23	1.259	0	0	0
1.260 —	33	1.261	1	33	33
1.262 —	13	1.263	2	26	52
1.264 —	6	1.265	3	18	54
1.266–1.268	1	1.267	4	4 (81)	16
	$\Sigma f = 100$			$\Sigma fu = 40$	$\Sigma fu^2 = 248$

$$\bar{u} = \frac{\Sigma fu}{\Sigma f} = \frac{40}{100} = 0.4$$

$$\overline{u^2} = \frac{\Sigma fu^2}{\Sigma f} = \frac{248}{100} = 2.48$$

$$Var(u) = \frac{\Sigma fu^2}{\Sigma f} - \left(\frac{\Sigma fu}{\Sigma f}\right)^2$$

$$= 2.48 - 0.16$$

$$= 2.32$$

So

$$S_u = \sqrt{2.32} = 1.523$$

Now

$$x = 0.002u + 1.259$$

So

$$\bar{x} = 0.002\bar{u} + 1.259$$

$$= 1.2598$$

$$\simeq 1.260$$

and

$$S_x = 0.002 \times S_u$$

$$= 0.002 \times 1.523$$

$$= 0.003046$$

$$\simeq 0.003$$

Notice in this case that it is rather more sensible to use coding than use a calculator directly. The reader might like to check this assertion.

Exercise 4.2

1. Find the standard deviation of the data given in
   (a) Exercise 2.1, question 1,
   (b) Exercise 2.2, question 1,
   (c) Exercise 2.3, questions 1 and 3.

2. Find the mean and variance of the integers 0–9 inclusive.

3. Using random number tables (p. 304) record the sum of four consecutive digits. Repeat this 100 times. Produce a grouped frequency distribution of the data and hence estimate the mean and variance. Compare these results with those obtained from question 2. Can you draw any conclusions?

4. Toss a dice 120 times, or use random number tables to simulate this experiment, and record the scores. Find the mean and variance of your data.

5. The number of goals scored per game by a particular football player during a recent season were as follows.

Number of goals	0	1	2	3	4 or more
Number of games	24	12	3	1	0

Calculate the mean and variance of the number of goals per game.

6. Two dice were thrown 100 times and in each case the total score was recorded.

Total score	2	3	4	5	6	7	8	9	10	11	12
Frequency	3	4	7	13	17	15	13	10	10	6	2

Find the mean and standard deviation of the total score.

7. The weekly pocket money of a sample of young pupils is recorded in the table below

Weekly pocket money (p)	Number of pupils
10 —	4
20 —	7
30 —	8
40 —	5
50–60	1

(a) Estimate the mean and standard deviation of the above distribution.
(b) If the pocket money of each child is increased by 20p write down the new mean and standard deviation.
(c) What will be the new mean and standard deviation if every child receives a 20% increase?

8. The marks of 6 boys were 7, 6, 5, 8, 7, 9. The marks for 5 girls were found to have a mean score of 5 with a standard deviation of 2.
   Find the mean and standard deviation of
   (a) the marks of the 6 boys,
   (b) the marks of the whole group of boys and girls.

9. A set of n numbers has mean M_1 and standard deviation S_1 and a further set of n numbers has a mean M_2 with a standard deviation of S_2. Show that the standard deviation of the whole set of $2n$ numbers is:

$$\sqrt{\tfrac{1}{2}(S_1^2 + S_2^2) + \tfrac{1}{4}(M_1 - M_2)^2}$$

10. Find the standard deviation of the set of $N_1 + N_2$ numbers if the first N_1 numbers have a mean of M_1 and variance S_1^2 and the other N_2 numbers have a mean of M_2 with variance S_2^2.

11. Prove the formula $\Sigma(x - \bar{x})^2 = \Sigma x^2 - n\bar{x}^2$.
   In a Middle School there are 253 girls whose ages have a mean 11.8 yr and a standard deviation 1.7 yr. There are also 312 boys whose ages have a mean 12.3 yr and a standard deviation 1.9 yr. Calculate the mean and standard deviation of the ages of all the 565 pupils.

AEB 1976

12. In a fishing competition, the total catches of 50 anglers had masses (to the nearest 0.1 kg) as given in the following table.

Mass in kg	0–0.2	0.3–0.7	0.8–1.2	1.3–1.7	1.8–2.2	2.3–3.7	3.8–5.2
Frequency	8	8	12	8	8	4	2

Draw a histogram to represent the frequency distribution.

State formulae for the mean and standard deviation of a frequency distribution, explaining the symbols used.

Prepare a table for the given data, showing the mid-values of the class intervals and all the terms in the summations that have to be calculated to obtain the mean and variance. Calculate the mean, variance and standard deviation of the distribution.

Given the additional information that all eight anglers placed in the first class interval caught nothing at all, obtain a revised value for the mean.

JMB

4.4 The cumulative frequency polygon

The cumulative frequency polygon (CF polygon) was discussed in Section 2.6 where it was shown that the median could be found with ease.

The median is the value below which 50% of the data lies. The rth percentile is defined as the value below which r% of the data lies. Thus the median is the 50th percentile.

The 25th percentile is often called the *first quartile* Q_1 and the 75th percentile is called the *third quartile* Q_3. The second quartile Q_2 is then the median. See Fig. 4.1.

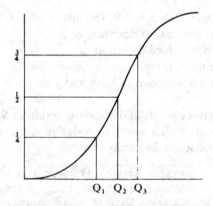

Fig. 4.1   A diagram to illustrate the positions of the 1st, 2nd and 3rd quartiles.

The value of $Q_3 - Q_1$ is called the *interquartile range* and $\frac{1}{2}(Q_3 - Q_1)$ is called the *semi-interquartile range*.

Since 50% of the data lies between the limits Q_1 and Q_3 a knowledge of the median and the semi-interquartile range gives a useful measure of spread. Thus if it were known that the median was 30 and $\frac{1}{2}(Q_3 - Q_1) = 4$, then it would be known that 50% of the data lies between 26 and 34.

This is a more useful measure of spread than the range in that it does not depend on extreme values. It also has the advantage of being obtained with ease.

Exercise 4.3

1. Return to Exercise 2.4 and for each question find the semi-interquartile range.

2. In an agricultural experiment the gains in mass, in kilograms, of 100 pigs during a certain period were recorded as follows.

Gain in mass (kg)	5–9	10–14	15–19	20–24	25–29	30–34
Frequency	2	29	37	16	14	2

Construct a histogram and a relative cumulative frequency polygon of these data. Obtain
(a) the median and the semi-interquartile range,
(b) the mean and the standard deviation.
Which of these pairs of statistics do you consider more appropriate in this case and why?

AEB 1977

3. The following table gives the maximum temperature in 60 cities on a certain day. Readings are to the nearest degree.

72 61 43 54 54 48 48 59 55 61 50 55
30 66 41 55 48 57 61 48 46 61 30 50
66 73 54 48 66 61 45 57 48 70 68 43
52 50 46 64 46 50 50 50 48 37 45 53
64 50 39 32 66 68 41 70 48 73 39 43

Gather these measurements into classes at intervals of 5°, starting at 29.5°.
From the information in this form construct a cumulative frequency table and draw a graph to display the cumulative frequency. Use this to find the median and semi-interquartile range.
For the grouped distribution calculate the mean and standard deviation, explaining what effect the grouping is likely to have had on each.

AEB 1977

4. The following table gives the age distribution of 3137 holders of a specific type of life assurance policy. (The true age of a policyholder whose nearest age is, for example, 44 lies between $43\frac{1}{2}$ and $44\frac{1}{2}$.)

Nearest age	21–25	26–30	31–35	36–40	41–45	46–50
Number	44	47	66	107	147	151
Nearest age	51–55	56–60	61–65	66–70	71–75	76–80
Number	208	305	396	488	471	358
Nearest age	81–85	86–90	91–95	96–100		
Number	222	104	18	5		

Estimate the mean age of a policyholder.
Calculate also the median age and the upper and lower quartiles.
Plot the data on a cumulative frequency diagram and from this estimate the median age and the upper and lower quartiles and compare with the values previously calculated.

MEI

Chapter 5 Probability laws

5.1 Introduction
In Chapter 3 the theory of probability was developed informally and the emphasis was on the development of skills. In this chapter the theory of probability will be developed more formally.

A sample space has been defined as the set of ALL possible outcomes of an experiment, game, result of counting etc. We can think of it as a set of points, each of which represents a possible outcome. See Fig. 5.1. An event E can now be thought of as a subset of the sample space. When an experiment is performed and the outcome is known, either the point representing the outcome is in E or it is not. In the former case the event E occurred and in the latter it did not.

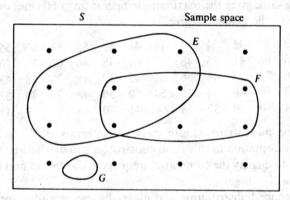

Fig. 5.1   Events E, F and G are subsets of the sample space S.

The above development suggests the use of Venn diagrams found in set theory. The notation of set theory is also very useful and some of the usual definitions are now introduced.

$E \cap F$	the set which contains those objects in both E and F	*read*, E intersection F
$E \cup F$	the set which contains those objects in either E or F (or both)	*read*, E union F
E'	the set which contains those objects not in E	*read*, the complement of E
$n(E)$	the number of objects in E	
$E \subset F$	this statement is true if every member of E is also a member of F	*read*, E is a subset of F

5.2 Equally likely sample spaces

Definition 5.1

> If S is a sample space and E is a possible event ($E \subset S$) then the probability that E occurs is
>
> $$P(E) = \frac{n(E)}{n(S)}$$

This definition agrees with the intuitive ideas developed in Chapter 3. Returning to Fig. 5.1 above that

$$P(E) = \frac{7}{16}$$

$$P(F) = \frac{6}{16}$$

$$P(S) = \frac{16}{16} = 1 \qquad \text{as this is certain to occur by the definition of a sample space}$$

$$P(G) = \frac{0}{16} = 0 \qquad \text{an event which cannot occur}$$

$$P(E \cap F) = \frac{3}{16}$$

$$P(E \cup F) = \frac{10}{16}$$

$$P(E') = \frac{9}{16}$$

The next task is to develop the algebra of probability and establish some of the laws of probability.

For the following, suppose E and F are any events.

The use of 'not'

$$P(\text{not } E) = P(E')$$

$$= \frac{n(E')}{n(S)}$$

$$= \frac{n(S) - n(E)}{n(S)}$$

$$= 1 - P(E)$$

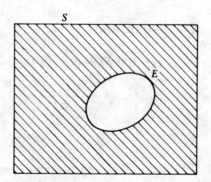

The use of 'or'

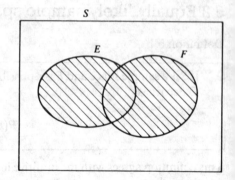

$$P(E \text{ or } F) = P(E \cup F)$$

$$= \frac{n(E \cup F)}{n(S)}$$

(a little
thought $= \dfrac{n(E) + n(F) - n(E \cap F)}{n(S)}$
shows)

$$= P(E) + P(F) - P(E \cap F)$$

Mutually exclusive events

Two events are said to be *mutually exclusive* if it is impossible for them both to occur simultaneously. In the language of set theory, the events E and F are mutually exclusive if $E \cap F = \emptyset$ and so $P(E \cap F) = 0$.

Thus if the events E and F are mutually exclusive

$$P(E \text{ or } F) = P(E) + P(F)$$

alternatively

$$P(E \cup F) = P(E) + P(F)$$

This result can be extended to

$$P(A_1 \text{ or } A_2 \text{ or } A_3 \text{ or } \dots) = P(A_1 \cup A_2 \cup A_3 \cup \dots)$$

$$= P(A_1) + P(A_2) + P(A_3) + \dots,$$

when the events $A_1, A_2, A_3, \dots$ are pairwise mutually exclusive.

The use of 'and'

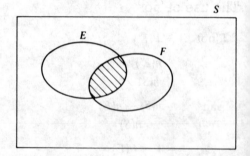

$$P(E \text{ and } F) = P(E \cap F)$$

$$= \frac{n(E \cap F)}{n(S)}$$

To simplify $P(E \cap F)$ further, a new concept is introduced.

Conditional probality

Definition 5.2

> The conditional probability of E given F written $P(E|F)$ is the probability that
> E occurs given that (we know) F occurs.

For example suppose a card is selected at random from a well shuffled pack. Let E be the event that the card is red and F be the event that the card is a heart.

Then $P(F) = \frac{1}{4}$ but $P(F|E) = \frac{1}{2}$. The probability that a heart is selected given that we know that the card is red is $\frac{1}{2}$ since of the 26 possible red cards 13 of them are hearts.

Again $P(E) = \frac{1}{2}$ but $P(E|F) = 1$ since if a heart has been selected it will obviously be red.

The alert reader will have noticed that some of the probabilities associated with the branches of a 'tree' diagram (introduced in Chapter 3) were conditional probabilities.

Definition 5.3

> A formal definition of the conditional probability of E given F is
> $$P(E|F) = \frac{P(E \cap F)}{P(F)} = \frac{P(E \text{ and } F)}{P(F)} \quad \text{providing } P(F) \neq 0$$

We justify this definition as follows.
If the event F has occurred then F is the new sample space.

$$P(E|F) = \frac{n(E \cap F)}{n(F)}$$

$$= \frac{n(E \cap F)}{n(F)} \times \frac{n(S)}{n(S)}$$

$$= \frac{n(E \cap F)}{n(S)} \times \frac{n(S)}{n(F)}$$

$$= \frac{n(E \cap F)/n(S)}{n(F)/n(S)}$$

$$= \frac{P(E \cap F)}{P(F)}$$

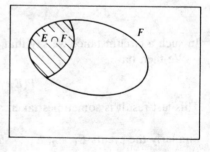

We can check this result by noting that if E is the event of obtaining a red card and F is the event of obtaining a heart mentioned above then

$$P(F|E) = \frac{P(F \cap F)}{P(E)} = \frac{\frac{1}{4}}{\frac{1}{2}} = \frac{1}{2}$$

which agrees with the result obtained earlier.

By rearranging the equation, we obtain

$$P(E \text{ and } F) = P(E \cap F)$$
$$= P(F) \times P(E|F)$$
$$= P(E) \times P(F|E) \text{ by symmetry (assuming}$$
$$P(E) \neq 0)$$

Independent events

Suppose that $P(E|F) = P(E)$, that is a knowledge that F occurs does not affect the probability that E occurs, than we can show that the reverse is true. That is the knowledge that E occurs does not effect the probability that F occurs.

For if

$$P(E|F) = P(E)$$

then

$$P(F|E) = \frac{P(F \cap E)}{P(E)}$$

$$= \frac{P(E|F) \times P(F)}{P(E)}$$

$$= \frac{P(E) \times P(F)}{P(E)}$$

$$= P(F)$$

In such a circumstance we say that the events E and F are *independent*.

We then have

$$P(E \cap F) = P(E) \times P(F)$$

This last result is sometimes taken as the definition of (statistical) independence.

Similarly the events $E_1, E_2, E_3, \ldots E_n$ are independent if

$$P(E_1 \cap E_2 \cap E_3 \cap \ldots \cap E_n) = P(E_1) \times P(E_2) \times P(E_3) \times \ldots \times P(E_n)$$

Definition 5.4

Two events E and F are said to be mutually exclusive if

$$P(E \cup F) = P(E) + P(F)$$

or equivalently

$$P(E \cap F) = 0$$

Two events E and F are said to be independent if

$$P(E \cap F) = P(E) . P(F)$$

or equivalently

$$P(E|F) = P(E) \quad \text{assuming } P(F) \neq 0$$

Summary

Statement 5.1

$$P(\text{not } E) = P(E')$$
$$= 1 - P(E)$$
$$P(E \text{ or } F) = P(E \cup F)$$
$$= P(E) + P(F) - P(E \cap F)$$

and if E and F are mutually exclusive then

$$P(E \cup F) = P(E) + P(F)$$
$$P(E \text{ and } F) = P(E \cap F)$$
$$= P(E) \times P(F|E)$$

and if E and F are independent then

$$P(E \cap F) = P(E) \times P(F)$$

5.3 Sample spaces

The probabilities associated with the outcomes contained within a sample space need not be equally likely.

For example, suppose the result of tossing a biased die can be modelled by the following sample space with associated probabilities as indicated.

Outcomes	1	2	3	4	5	6
Probabilities	$\frac{1}{10}$	$\frac{2}{10}$	$\frac{1}{10}$	$\frac{2}{10}$	$\frac{1}{10}$	$\frac{3}{10}$

Then the probability of the event E is defined by

$P(E)$ = sum of all the probabilities associated with the members of E.

Thus in the example above

$$P(\text{even}) = \frac{2}{10} + \frac{2}{10} + \frac{3}{10}$$

$$= 0.7$$

The results developed above still apply in this case though the proofs are a little more difficult.

Example 5.1

A card is chosen at random from an ordinary pack of cards. Write down the probability that the card is
(a) black or an ace
(b) black and an ace ($\equiv$ a black ace)
(c) neither black nor an ace

(a) $P(\text{black or an ace}) = P(\text{black}) + P(\text{ace}) - P(\text{black and an ace})$

$$= \frac{26}{52} + \frac{4}{52} - \frac{2}{52}$$

$$= \frac{28}{52}$$

$$= \frac{7}{13}$$

(b) $P(\text{black and an ace}) = P(\text{black}) \cdot P(\text{ace|black})$

$$= \frac{26}{52} \times \frac{2}{26}$$

$$= \frac{2}{52}$$

OR

$$= P(\text{ace}) \cdot P(\text{black|ace})$$

$$= \frac{4}{52} \times \frac{2}{4}$$

$$= \frac{2}{52}$$

(c) P(neither black nor an ace) = 1 − P(black or an ace)

$$= 1 - \frac{28}{52}$$

$$= \frac{24}{52}$$

$$= \frac{6}{13}$$

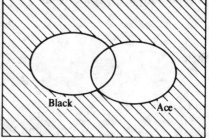

Which agrees with the results obtained by using methods introduced earlier.

Example 5.2 Justification of the 'tree-diagram' method

Two balls are taken out at random from a bag containing 6 red and 2 blue balls. Find:
(a) P(RR)
(b) P(just one B)
(c) P(at least one B)

(a) P(R, R) = P(R first and R second)

$$= P(R \text{ first}) \cdot P(R \text{ second}|R \text{ first})$$

$$= \frac{6}{8} \times \frac{5}{7} = \frac{15}{28}$$

and this reflects the method indicated by a 'tree diagram'.

Notice how easy the conditional probabilities are to compute.

(b) P(just one B) = P(B first and R second OR R first and B second)

$$= P(B \text{ first and } R \text{ second}) + P(R \text{ first and } B \text{ second})$$

since the events are mutually exclusive

$$= P(B \text{ first}) \cdot P(R \text{ second}|B \text{ first}) + P(R \text{ first}) \cdot P(B \text{ second}|R \text{ first})$$

$$= \frac{2}{8} \times \frac{6}{7} + \frac{6}{8} \times \frac{2}{7} = \frac{3}{7}$$

Again this matches precisely the method indicated by a 'tree-diagram'. In practice there is no need to write as much as we have above.

(c) P(at least one B) = 1 − P(no B)

$$= 1 - P(R, R)$$

$$= 1 - \frac{15}{28}$$

$$= \frac{13}{28}$$

In general suppose a 'branch' of a 'tree' consists of events A and B with associated probabilities $p_1\ p_2$ respectively

$$\underline{\quad p_1 \quad} A \underline{\quad p_2 \quad} B$$

then

$$p_1 = P(A) \quad p_2 = P(B|A)$$

and

$$P(A \text{ and then } B) = P(A) \cdot P(B|A)$$

$$= p_1\, p_2$$

By the nature of a 'tree' the events corresponding to 'branch' π_1 and a 'branch' π_2 are mutually exclusive and so

$$P(\pi_1 \text{ or } \pi_2) = P(\pi_1) + P(\pi_2).$$

Example 5.3
The events A and B are such that

$$P(A) = \tfrac{1}{3}$$

$$P(A \text{ or } B \text{ but not both } A \text{ and } B) = \tfrac{5}{12}$$

$$P(B) = \tfrac{1}{4}.$$

Calculate $P(A \cap B)$, $P(A' \cap B)$, $P(A|B)$ and $P(B|A')$.

The event "A or B but not both A and B" is represented by the shaded region in the diagram.

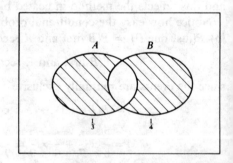

Since

$$P(A \text{ or } B) = P(A) + P(B) - P(A \text{ and } B)$$

$$P(A \text{ or } B \text{ but not both } A \text{ and } B) = P(A) + P(B) - 2P(A \text{ and } B).$$

Thus

$$\frac{5}{12} = \frac{1}{3} + \frac{1}{4} - 2\,P(A \text{ and } B)$$

$$\Rightarrow \quad P(A \text{ and } B) = \frac{1}{12} = P(A \cap B)$$

$$P(A' \cap B) = P(B) - P(A \cap B) \qquad \text{as can be seen from the diagram}$$

$$= \frac{1}{4} - \frac{1}{12}$$

$$= \frac{1}{6}$$

$$P(A|B) = \frac{P(A \text{ and } B)}{P(B)}$$

$$= \frac{1}{12} \bigg/ \frac{1}{4}$$

$$= \frac{1}{3}$$

$$P(B|A') = \frac{P(B \text{ and } A')}{P(A')}$$

$$= \frac{1}{6} \bigg/ \frac{2}{3}$$

$$= \frac{1}{4}$$

Example 5.4

Three bags each contain a number of red, white and blue balls as indicated in the diagram.

	Bag 1	Bag 2	Bag 3
Red	1	4	2
White	2	2	2
Blue	3	4	2

(a) If a ball is picked at random find, assuming the obvious notation
   (i) $P(\text{Bag 2})$
   (ii) $P(R|\text{Bag 2})$
   (iii) $P(R \text{ and Bag 2})$
   (iv) $P(R)$
   (v) $P(\text{Bag 2}|R)$

(b) If a bag is picked at random and a ball is selected at random from this bag calculate the above probabilities again.

(a)
   (i) $P(\text{Bag 2}) = \frac{10}{22} = \frac{5}{11}$

(ii) $P(R|\text{Bag 2}) = \dfrac{4}{10} = \dfrac{2}{5}$ (iv) $P(R) = \dfrac{7}{22}$

(iii) $P(R \text{ and Bag 2}) = \dfrac{4}{22} = \dfrac{2}{11}$ (v) $P(\text{Bag 2}|R) = \dfrac{4}{7}$

The above were obtained by inspection of the diagram.

(b)

(i) $P(\text{Bag 2}) = \dfrac{1}{3}$ chosen at random

(ii) $P(R|\text{Bag 2}) = \dfrac{4}{10} = \dfrac{2}{5}$

(iii) $P(R \text{ and Bag 2}) = P(\text{Bag 2})\, P(R|\text{Bag 2})$

$$= \frac{1}{3} \times \frac{2}{5}$$

$$= \frac{2}{15}$$

(iv) $P(R) = P(\text{Bag 1 and } R) + P(\text{Bag 2 and } R) + P(\text{Bag 3 and } R)$ (since the events are mutually exclusive)

$= P(\text{Bag 1})\, P(R|\text{Bag 1}) + P(\text{Bag 2})\, P(R|\text{Bag 2}) + P(\text{Bag 2})\, P(R|\text{Bag 3})$

$$= \frac{1}{3}\cdot\frac{1}{6} + \frac{1}{3}\cdot\frac{4}{10} + \frac{1}{3}\cdot\frac{2}{6}$$

$$= \frac{5 + 12 + 10}{90}$$

$$= \frac{27}{90}$$

$$= \frac{3}{10}$$

(v) $P(\text{Bag 2}|R) = \dfrac{P(R \text{ and Bag 2})}{P(R)}$ by definition.

$$= \frac{\dfrac{2}{15}}{\dfrac{3}{10}}$$

$$= \frac{4}{9}$$

Exercise 5.1

1. State which of the following pairs of events are independent and which are mutually exclusive.

   (a) An unbiased coin is tossed twice.

   E_1 represents the event that two heads are obtained.

   E_2 represents the event that a head is obtained on the first throw.

   (b) A card is selected at random from a full pack of cards.

   F_1 represents the event of obtaining a red card.

   F_2 represents the event of obtaining an ace.

   (c) An unbiased die is tossed twice and the total score noted.

   G_1 represents the event that a score of 11 is obtained.

   G_2 represents the event that a 3 is obtained on the first die.

2. A dart is thrown at a board and is equally likely to land in any one of eight squares numbered 1 to 8 inclusive.

   A represents the event that the dart lands in square 5 or 8.

   B represents the event that it lands in square 2, 3 or 4.

   C represents the event that it lands in square 1, 2, 5 or 6.

   Find (i) $P(A \cap B)$ (ii) $P(A \cap C)$ (iii) $P(B \cap C)$

   (iv) $P(A|B)$ (v) $P(B|C)$ (vi) $P(C|A)$

   (a) Which two events are mutually exclusive?
   (b) Which two events are statistically independent?

3. A motorist makes a journey to work which involves crossing two traffic lights A and B. He reckons that the probabilities of being delayed are independently 0.4, and 0.3 respectively. Find the probability of there being 0, 1, 2, 3 delays on the journey.

4. A fruit machine has three windows, in each of which appears independently one of five different pictures. If there is a one-fifth chance of obtaining a cherry in each window, find the probability of obtaining
   (a) three cherries,
   (b) one cherry in the first window,
   (c) at least one cherry.

5. In a mixed bag of nuts there are a large number of small nuts and twice as many large ones. It is known that 10% of the large nuts and 5% of the small ones are bad. Find the probability that
   (a) the first nut chosen at random is
      (i) large,
      (ii) small and bad,
      (iii) bad:

(b) the first two chosen at random are
  (i) small,
  (ii) edible.

6. Three numbers are chosen independently from the digits 0, 1, 2, ..., 9. Find the probability that
(a) all the numbers x satisfy $x \leqslant 3$,
(b) all the numbers are distinct,
(c) all the numbers are the same.

7. The chance of any of three runners in a race breaking a record are 1/3, 3/4 and 1/2 respectively. What is the probability that the record will be broken if it is assumed that the above probabilities are independent of one another? Is the assumption of independence realistic?

8. The probability that two independent events A and B will occur are p_1 and p_2 respectively. Find the probability that
(a) both A and B occur,
(b) A occurs but B does not,
(c) just one of them occurs,
(d) neither A nor B occurs.

9. A number is chosen from the first 100 positive integers. If each is equally likely to be selected, calculate the probability that the integer is
(a) divisible by 5,
(b) divisible by 3,
(c) divisible by 15,
(d) neither divisible by 4 nor 5.

10. A garden has two flower beds. The first bed has 10 daffodils and 20 tulips and the second has 20 daffodils, 10 tulips and 30 snowdrops. A flower is chosen at random. Find the probability that
(a) it is a tulip,
(b) it is a daffodil from the second bed,
(c) it is from the first bed,
(d) given it came from the second bed it was a daffodil,
(e) given it was a daffodil it came from the second bed.

11. Write down two events which, if possible, are
(a) independent,
(b) mutually exclusive,
(c) mutually exclusive but not independent,
(d) independent but not mutually exclusive,
(e) independent and mutually exclusive.

12. An air-to-air missile has 7/8 probability of hitting a target. If five missiles are launched, what is the probability that the target is not destroyed.

13. A relative frequency distribution is obtained from a frequency distribution by dividing each of the frequencies in the distribution by the total number of observations recorded in the table (i.e. Σf).

The relative frequencies are, of course, approximations to the probabilities associated with certain outcomes.

Return to Exercise 2.3 question (1) and draw up a relative frequency distribution of the data. Estimate the probability that a car chosen at random will be travelling
   (a) within the speed limit of 30 mph
   (b) between 50 mph and 60 mph.

14. Toss ten coins 100 times and in each case record the number of heads.

Produce a relative frequency distribution of the data. Draw a histogram and describe the shape of the distribution in words.

Estimate the probability of obtaining
   (a) 0 heads,
   (b) 1 head.
Check your results by calculation, assuming the coins to be unbiased. Explain any discrepancy in your results.

15. A card is selected at random from a full pack of cards. If E, F and G are respectively the events of obtaining a heart, a red and an ace find
   (a) $P(F|E)$,
   (b) $P(F|G)$,
   (c) $P(G|E)$.
Show that E and G are independent, G and F are independent, but F and E are not.

16. A and B are two events.
   (a) If A and B are mutually exclusive under what condition can they also be independent?
   (b) If A and B are independent under what conditions can they also be mutually exclusive?
   (c) Under what conditions will $P(A|B) \neq P(B|A)$?

17. If the probability that a male lives to age 40 is 0.9 and that he lives to age 60 is 0.7 what is the probability that if he reaches the age of 40 years he will live to age 60?

18. A box contains two balls. It is equally likely that the box contains 0, 1 or 2 black balls. A ball is drawn at random and found to be black. What is the probability that the other is black?

19. Suppose I had to find three light bulbs from a collection of 10, all but 4 of which are satisfactory. If I test each one in turn find the probability that I find three working bulbs
   (a) among the first five bulbs tested
   (b) when five bulbs have been tested, but not before

20. In a game of Russian Roulette a gun has a bullet in one of its six chambers; the

chamber is spun and the gun is fired. A player decides to have two trials (at most!).
Find the probability that he is still alive after the two trials
   (a) if he spins the chamber between trials
   (b) if he does not spin the chamber between trials

21. Two events A and B are such that $P(A) = 1/4$ $P(B) = 1/2$ and $P(A$ and $B) = 1/8$.
Evaluate
   (a) $P(A \cup B)$,
   (b) $P((A \cap B)')$,
   (c) $P(A' \cap B)$,
   (d) P(neither A nor B occurs),
   (e) P(either A or B but not both occur).

22. If $P(A) = \dfrac{1}{4}$, $P(B) = \dfrac{1}{3}$, $P(A \cup B) = \dfrac{1}{2}$, find:

   (a) $P(A \cap B)$,   (b) $P(A|B)$,   (c) $P(B|A)$,   (d) $P(A'|B)$,   (e) $P(A|B')$.

23. Three cards are drawn at random without replacement from a pack of ten cards
which are numbered from 1 to 10, respectively. Calculate
   (a) the probability that the numbers drawn consist of two even numbers and one
       odd number,
   (b) the probability that at least one of the numbers drawn is a perfect square
       greater than 1,
   (c) the probability that the smallest number drawn is the 5.
                                                                                    JMB
24. (a) Two cards are drawn at random, without replacement, from an ordinary pack
       of 52 cards. Find the probability that they are:
       (i) of the same suit,
       (ii) of the same value (both aces, both kings, etc.),
       (iii) either of the same suit or the same value.
   (b) Two cards are drawn at random, one from each of two ordinary packs. Find
       the probability that they are
       (i) of the same suit,
       (ii) of the same value,
       (iii) either of the same suit, the same value, or both.
   (c) Three fair cubical dice are thrown. Find the probability that the sum of the
       numbers of spots on the upper faces is a perfect square.
                                                                                    AEB
25. In Baltravia there are just three shipping companies A, B and C. These companies
own ships of three types: oil tankers, cargo ships and passenger ships. The
compositions of their fleets are given in the table below.

Company	Oil tankers	Cargo ships	Passenger ships
A	80	20	20
B	40	40	20
C	10	30	40

(a) Determine the probability that a randomly selected Baltravian ship is owned by company A.

(b) Determine the probability that a randomly selected Baltravian oil tanker is owned by company A.

(c) One ship is selected at random from the fleet of each of the three companies. Determine the probability that the three ships selected are
 (i) all tankers,
 (ii) all of the same type,
 (iii) all tankers, if it is known that they are all of the same type.

CAMB

26. In a class of 30 pupils, 12 walk to school, 10 travel by bus, 6 cycle and 2 travel by car. If four pupils are picked at random, obtain the probabilities that (i) they all travel by bus, (ii) they all travel by the same means.

If two are picked at random from the class, find the probability that they travel by different means.

In picking out pupils from the class, find the probability that more than three trials are necessary before a pupil who walks to school is selected.

JMB

Chapter 6   Expectation algebra

6.1 Introduction

So far we have considered the collection of real data, perhaps simplified into frequency distributions, and the calculation of various 'statistics'. The term 'statistic' means any number description of the data, such as the mean, mode, median, range, variance etc. We have also developed the theory of probability, where we noted the use of frequency distributions to estimate probabilities – by the corresponding relative frequency.

The link can be taken much further as will be seen.

A die was thrown sixty times and the results obtained are shown in the following frequency distribution.

Score	x	1	2	3	4	5	6
Frequency	f	9	8	11	14	9	9

We can use this frequency distribution to calculate any statistic of interest, such as the mean.

$$\bar{x} = \frac{\Sigma fx}{\Sigma f} = \frac{\Sigma fx}{N}, \qquad \text{where } N = \Sigma f = 60$$

$$= \frac{1}{60}(9 \times 1 + 8 \times 2 + 11 \times 3 + 14 \times 4 + 9 \times 5 + 9 \times 6)$$

If desired this could be written as

$$\bar{x} = \Sigma \frac{fx}{N} = \frac{9}{60} \times 1 + \frac{8}{60} \times 2 + \frac{11}{60} \times 3 + \frac{14}{60} \times 4 + \frac{9}{60} \times 5 + \frac{9}{60} \times 6,$$

where the fractions f/N represent the relative frequency with which the various values occur. These represent our best estimates of the probability that particular scores occur – given that we have no further information regarding the die, such as its being fair.

We would be much happier estimating probabilities using relative frequencies if we tossed the die rather more than sixty times, since we would argue that as

$$N \to \infty, \frac{f}{N} \to \text{the probability } p_r$$

NB. The notation $\to$ is read 'tends to' and implies a limiting process. As N gets larger so the relative frequency f/N gets closer to the (true) probability.

6.2 Expectation and random variables

If the mean $\bar{x}$ is calculated from the above distribution it will be found to be 3.55. If we continued tossing the die, recording the results, and calculated the mean at various intervals we might obtain say:

$$\bar{x} = 3.55, \quad \bar{x} = 3.52, \quad \bar{x} = 3.506, \quad \bar{x} = 3.5002$$

and wonder whether there was a limit to which the $\bar{x}$'s were tending.

Using the fact that $\bar{x} = \Sigma(f/N)x$ and that as $N \to \infty, f/N \to p$, we expect that

$$\bar{x} \to \Sigma p_r x = P(1) \times 1 + P(2) \times 2 + P(3) \times 3 + P(4) \times 4 + P(5) \times 5 + P(6) \times 6$$

This leads us to introduce the following ideas.

Random variables

Suppose a variable can be associated with the outcomes of an experiment, so that each outcome establishes a value of the variable. For example we may toss a coin five times, and count the number of heads obtained, or locate a family and count the number of children. In this way particular probabilities can be assigned to the values that X can assume. Such a variable is called a *variate* or *random variable*.

> If X can take values $x_1, x_2 \ldots x_n$ each with probability $p_1, p_2 \ldots p_n$, then such a table represents a *probability distribution* for X.

It is important to realise that any table consisting of a sequence of numbers $x_1, x_2, \ldots x_n$ each with associated numbers $p_1, p_2, \ldots p_n$ is a probability distribution providing only that

(i) each $p_r \geqslant 0$

(ii) $\Sigma p_r = 1$

Notice that capital letters X, Y and Z are reserved for variates and the corresponding lower case letters x, x_1, y, y_1 for values the variate can assume.

Expectation

If X is a variate we define the expected value of X as

$$E(X) = \Sigma p_r x$$

In words: the expected value of X (the expectation of X) is found by multiplying the values that X can take with the probabilities that it takes these values and adding the results together.

In summary

Definition 6.1

> A variable X whose value is determined by the outcome of a chance experiment is called a *variate*.
>
> A table or function showing the values that X can assume together with the associated probabilities with which it takes these values is called a probability distribution for X.
>
> If X is a variate then the expectation of X written $E(X)$ is defined by
>
> $$E(X) = \Sigma p_r x$$
>
> where the summation is taken over all the possible values that X can assume. Each such value is multiplied by the probability with which it can assume that value.

Example 6.1

Suppose a fair die is tossed and X represents the scores obtained, find the $E(X)$,

X	1	2	3	4	5	6
p_r	1/6	1/6	1/6	1/6	1/6	1/6

It is because the die is fair that we can fill the probabilities in the table with ease.

$$E(X) = \Sigma p_r x$$

$$= \frac{1}{6} \times 1 + \frac{1}{6} \times 2 + \frac{1}{6} \times 3 + \frac{1}{6} \times 4 + \frac{1}{6} \times 5 + \frac{1}{6} \times 6$$

$$= \frac{1}{6}(1 + 2 + 3 + 4 + 5 + 6)$$

$$= \frac{21}{6}$$

$$= 3.5$$

To say that $E(X) = 3.5$ is to suggest that if we tossed a fair die many times then the mean score would be very close to 3.5. Or more generally as

$$N \to \infty, \quad \bar{x} \to E(X)$$

Thus whilst $\bar{x}$ represents the actual mean obtained, $E(X)$ represents the theoretical mean. We shall often refer to the mean of a distribution, and allow the context to indicate whether we are referring to the actual mean of observed data or the expected value.

Definition 6.2

> In general if $g(X)$ is any function of X then we define
> $$E(g(X)) = \Sigma p_r g(x)$$

Using the frequency distribution introduced above we could also calculate the variance of the scores obtained using the formula

$$\text{Var}(x) = \frac{\Sigma f x^2}{n} - \left(\frac{\Sigma f x}{n}\right)^2 \qquad \text{where } n = \Sigma f$$

$$= \Sigma\left(\frac{f}{n}\right)x^2 - \left[\Sigma\left(\frac{f}{n}\right)x\right]^2$$

and as $n \to \infty$ this expression tends to

$$\to \Sigma p_r x^2 - (p_r x)^2$$

So we define the variance of the variate X by

Definition 6.3

$$Var (X) = E(X^2) - E(X)^2$$

or equivalently and more usually

$$Var (X) = E[(X - E(X))^2]$$

Example 6.2

Suppose the variate X represents the score obtained from tossing a fair die. Find the value of Var (X).

We have found $E(X) = 3.5$. To find $E(X^2)$ produce a new distribution

X^2	1	4	9	16	25	36
p_r	1/6	1/6	1/6	1/6	1/6	1/6

Notice that the probabilities are the ones corresponding to the underlying X values.

$$E(X^2) = \Sigma p_r x^2$$

$$= \tfrac{1}{6}(1 + 4 + 9 + 16 + 25 + 36)$$

$$= \frac{91}{6}$$

$$= 15\tfrac{1}{6}$$

$$Var (X) = E(X^2) - E(X)^2$$

$$= 15\tfrac{1}{6} - 12\tfrac{1}{4}$$

$$= \frac{35}{12}$$

$$\simeq 2.92$$

This theoretical variance should be compared with the actual variance obtained for Exercise 4.2, question 4.

6.3 The uniform (or rectangular) distribution

Definition 6.4

A variate follows the (discrete) uniform distribution if its probability distribution is of the form

$$P(X = x) = \frac{1}{n}, \quad \text{when } x = 1, 2, 3, \ldots, n$$

$$= 0 \quad \text{otherwise.}$$

In this case $E(X) = \dfrac{n + 1}{2}, \quad Var (X) = \dfrac{n^2 - 1}{12}$

This common distribution was met in the previous example where an unbiased die was considered.

Using the above results when $n = 6$ we find

$$E(X) = \frac{7}{2} \qquad\qquad Var\ (X) = \frac{35}{12}$$

$$= 3.5 \qquad\qquad\qquad \simeq 2.92$$

Which agrees with the results found directly.
To prove these results we require two formulae:

(a) $1 + 2 + 3 + 4 \ldots + n = \dfrac{n}{2}(n + 1)$,

(b) $1^2 + 2^2 + 3^2 + \ldots + n^2 = \dfrac{n}{6}(n + 1)(2n + 1)$.

Now

$$E(X) = \frac{1}{n} \times 1 + \frac{1}{n} \times 2 + \frac{1}{n} \times 3 + \ldots + \frac{1}{n} \times n$$

$$= \frac{1}{n}(1 + 2 + 3 + \ldots + n)$$

$$= \frac{1}{n}\left[\frac{n}{2}(n + 1)\right]$$

$$= \frac{n + 1}{2}$$

and

$$E(X^2) = \frac{1}{n}(1^2 + 2^2 + \ldots + n^2)$$

$$= \frac{1}{n}\left[\frac{n}{6}(n + 1)(2n + 1)\right]$$

$$= \frac{1}{6}(n + 1)(2n + 1)$$

But

$$Var\ (X) = E(X^2) - E(X)^2$$

$$= \frac{1}{6}(n + 1)(2n + 1) - \frac{(n + 1)^2}{4}$$

$$= \frac{n + 1}{12}[2(2n + 1) - 3(n + 1)]$$

$$= \frac{n + 1}{12}(n - 1)$$

$$= \frac{n^2 - 1}{12}$$

Exercise 6.1

1. Three values of a random variable, together with their associated probabilities are given below.
   Calculate the expectation of the variate.

(a) X	0	1	2	3	4	5
p_r	1/15	2/15	1/3	1/5	2/15	2/15

(b) X	−2	−1	0	1	2
p_r	2/10	1/10	2/10	3/10	2/10

(c) X	1	2	3	4	5	6
p_r	1/3	0	1/6	0	1/2	0

   A new random variable Y is constructed so that $Y = 2X - 1$.
   Draw tables representing the probability distribution for Y in each of the above and in each case calculate $E(Y)$.
   Similarly find
   (i) $E(Z)$ where $Z = 3X$
   (ii) $E(W)$ where $W = 2Y$
   (iii) $E(V)$ where $V = X^2$

2. If the variate X has a probability distribution

X	−1	2	4	8
p_r	1/10	2/5	3/10	1/5

   (a) Find $E(X)$, $E(X^2)$ and Var (X)
   (b) If $Y = 3X - 6$ find $E(Y)$ and Var (Y)

3. A game consists of tossing two unbiased coins. If the number of pennies won on any play of the game is equal to three times the number of heads showing what are the average winnings (a) for one game, (b) for 10 games?
   How much should it cost in order for the game to be fair to both players?
   Hint: let X be the variate representing the possible winnings, tabulate its probability distribution and hence calculate $E(X)$. Subtract the amount you think it is fair to pay from each of the values the variate X can assume to produce a new probability distribution. Show directly that the expectation of this new distribution is zero.

4. Two dice are thrown together and the total score X is noted. If a player wins 30p for a score of 11 or 12 and 10p for a score of 6 or below, but has to pay 10p otherwise, find his average winnings after 20 games. State any assumptions you have made.

5. A die is designed so that the probability of its landing on any side is as shown below. Find the theoretical mean and variance of the score obtained when the die is tossed.

1	2	3	4	5	6
1/8	2/8	1/8	2/8	1/8	1/8

6. A man places three £1 notes and one £5 note into a bag. If he takes (a) 1 note, (b) 2 notes out of the bag at random find in each case the expected amount and the variance of the amount taken. Explain in your own words, in some detail, what these answers mean to you.

7. A man pays £1 to play a game with a die, where a score of 6 with one toss wins £3 and a score of 1 wins £2. Find his expected winnings if the stake is never returned.

6.4 Expectation algebra

In Exercise 6.1, question 1 it may have been noticed that when $Y = 2X - 1$

$$E(Y) = 2E(X) - 1$$

If this result had been used then, it would have made the calculations less tedious.

Statement 6.1

If $Y = aX + b$,   where a, b are constants and X and Y are variates

then

$$E(Y) = aE(X) + b$$

Proof:

$$E(Y) = \Sigma p_r y$$
$$= \Sigma p_r(ax + b)$$
$$= \Sigma p_r ax + \Sigma p_r b$$
$$= a\Sigma p_r x + b\Sigma p_r$$
$$= aE(X) + b, \quad \text{since } \Sigma p_r = 1 \text{ in a probability distribution.}$$

Statement 6.2

If X is a variate and a and b are constants then

(i) $E(aX) = aE(X)$

(ii) $E(b) = b$

These follow from Statement 6, in the first case by letting $b = 0$ and in the second by letting $a = 0$.

Example 6.3
If the variate X has mean μ and variance σ^2 then find the mean of Y and the variance of Z where $Y = 4X - 1$ and $Z = 2X$.
Since

$$Y = 4X - 1$$

$$E(Y) = 4E(X) - 1$$

$$= 4\mu - 1$$

$$\text{Var }(Z) = \text{Var }(2X)$$

$$= E(4X^2) - E(2X)^2, \qquad \text{by definition}$$

$$= 4E(X^2) - (2E(X))^2, \qquad \text{by Statement 6.2(i).}$$

$$= 4(E(X^2) - E(X)^2)$$

$$= 4 \text{ Var }(X)$$

$$= 4\sigma^2$$

Example 6.4
Show that if $Y = 3X^2 + 2X + 1$ where X is a variate with mean μ and variance σ^2 then

$$E(Y) = 3\sigma^2 + 3\mu^2 + 2\mu + 1.$$

We have no results to use and so return to the basic definitions.

$$E(Y) = \Sigma p_r y$$

$$= \Sigma p_r (3x^2 + 2x + 1)$$

$$= \Sigma p_r 3x^2 + \Sigma p_r 2x + \Sigma p_r$$

$$= 3\Sigma p_r x^2 + 2\Sigma p_r x + \Sigma p_r$$

$$= 3E(X^2) + 2E(X) + 1$$

$$= 3(\sigma^2 + \mu^2) + 2\mu + 1$$

The fact that $E(X^2) = \sigma^2 + \mu^2$ is worth remembering, it follows directly from

$$\text{Var }(X) = \sigma^2 = E(X^2) - \mu^2.$$

Statement 6.3

> If $g(X)$ and $h(X)$ are any functions of the variate X then
> $$E(g(X) + h(X)) = E(g(X)) + E(h(X))$$

No proof.

Exercise 6.2

1. If X is a random variable with mean μ and variance σ^2, simplify in terms of μ and σ^2

   (a) $E(3X)$, (b) $E(2X + 1)$, (c) $E(X^2)$,
   (d) $E(3X^2 + 5X + 9)$, (e) Var $(5X)$, (f) Var $(3X - 2)$,
   (g) Var $(4X + 6)$, (h) $E(X^2 + 2X)$.

2. If a and b are constants show that

   $$\text{Var } (aX + b) = a^2 \text{ Var } (X)$$

   and deduce that

   (i) Var $(aX) = a^2$ Var (X),
   (ii) Var $(b) = 0$.

3. If Y is a random variable with mean 2μ and variance σ^2, simplify in terms of μ and σ^2

   (a) $E(5Y)$ (b) $E(2Y - 6)$
   (c) Var $(4Y)$ (d) Var (9)
   (e) Var $(3Y - 6)$ (f) Var $(-Y)$
   (g) Var $(6 - Y)$ (h) Var $(4 - 2Y)$

4. Draw suitable diagrams to explain the results of 3(c), (d), (f).

5. Given that $P(X = r) = A(\frac{1}{2})^r$ for $r = 1, 2, 3, 4, 5$ where A is a constant, find the value of A and hence establish $P(X \leq 2)$ and $E(X)$.

6. A boy spins a coin up to four times until a head appears. Construct the probability distribution for the variate X which represents the number of times the coin is tossed and find the expected number of times the coin is tossed.

7. Any natural number can be reduced to a digit in the following way.

   $$31 \longrightarrow 3 + 1 \longrightarrow 4$$

   $$69 \longrightarrow 6 + 9 \longrightarrow 15 \longrightarrow 1 + 5 \longrightarrow 6$$

   If a natural number X such that $20 \leq X \leq 40$ is selected at random find the mean and variance of the reduced number associated with X.

Chapter 7   The Binomial and Geometric distributions

7.1 The derivation of the Binomial distribution

In this chapter one of the most useful probability distributions encountered in statistics is introduced. It can be used to model a certain class of situations which frequently occur in statistical investigations.

If a coin is tossed five times we might be interested in establishing the probability of obtaining exactly four heads. Again if a die is tossed eight times we may be asked to find the probability of obtaining exactly three sixes. Suppose a machine produces components which may or may not be defective, what is the probability that none of four selected at random are defective?

Common to each of these situations is the requirement to find the probability that a particular event will occur exactly r times when the experiment is repeated n times, under exactly similar conditions. We are interested in whether the event occurs or not rather than in the particular event. It is common practice to claim a 'success' when the event occurs and a 'failure' when it does not. Though this is not to imply that a success is considered desirable. In the examples above a success is obtained when either a head, six or defective component is observed after a particular experiment or trial is performed.

In addition, in order to ensure that the experiments are repeated under exactly similar conditions we shall require that the trials or experiments are independent of one another. This is not to suppose that the coins or die are necessarily fair but rather the result of one trial is unaffected by the result of previous trials. It is not obvious that the machine mentioned above satisfies these conditions since it is conceivable that once the machine starts to produce defective items it may continue to do so until it is repaired. It must be stressed that our model will only be satisfactory if the condition of independence between trials can be assumed true.

We consider now the coin tossing experiment mentioned above in more detail. Suppose the coin has a probability of 1/3 that it lands head side up when tossed. Suppose further that the coin is tossed five times, we find

(a) $P(\text{TTTTH})$
(b) $P(\text{exactly one 'head' is obtained}) = P(1\text{H})$
(c) $P(\text{exactly three 'heads' are obtained}) = P(3\text{H})$

In this example $P(\text{H}) = 1/3$ so $P(\text{T}) = 2/3$ and we shall consider a 'head' to be a success.

(a) Now the event TTTTH consists of four failures followed by a success, and because each toss is independent the multiplication law introduced in Chapter 5 gives

$$P(\text{TTTTH}) = P(\text{T}) \times P(\text{T}) \times P(\text{T}) \times P(\text{T}) \times P(\text{H})$$

$$= \frac{2}{3} \times \frac{2}{3} \times \frac{2}{3} \times \frac{2}{3} \times \frac{1}{3}$$

$$= \frac{16}{243}$$

$$\simeq 0.066$$

It can readily be seen that the order of the sequence of successes and failures is immaterial, so for example

$$P(\text{HTTTT}) = P(\text{THTTT}) = P(\text{TTTTH})$$

(b) In order to obtain exactly one 'head' in five tosses the only possible events are

TTTTH or TTTHT or TTHTT or THTTT or HTTTT

and these are mutually exclusive so

$$P(\text{exactly one 'head' is obtained}) = P(1\text{H})$$

$$= P(\text{TTTTH}) + P(\text{TTTHT}) + P(\text{TTHTT}) + P(\text{THTTT}) + P(\text{HTTTT})$$

$$= 5 \left(\frac{2}{3}\right)^4 \frac{1}{3}$$

$$\simeq 0.33$$

(c) To find the probability of obtaining exactly three 'heads' when the coin is tossed five times, note that

$$P(\text{TTHHH}) = P(\text{T}) \times P(\text{T}) \times P(\text{H}) \times P(\text{H}) \times P(\text{H})$$

$$= \frac{2}{3} \times \frac{2}{3} \times \frac{1}{3} \times \frac{1}{3} \times \frac{1}{3}$$

$$= \left(\frac{2}{3}\right)^2 \left(\frac{1}{3}\right)^3$$

$$= \frac{4}{243}$$

$$\simeq 0.016$$

Now since the arrangement of the three 'heads' in the sequence does not alter the probability we need only establish the number of such sequences. But this is merely $^5C_3$, the number of ways of choosing three objects from five. To see that this is so imagine five spaces

— — — — —

We choose three of the five spaces, (we can do this in $^5C_3$ ways) and put 'H' in these spaces, and then place 'T' in the other two spaces.

Thus $P(\text{exactly three 'heads'}) = P(3H)$

$$= {}^5C_3 \times P(\text{TTHHH})$$

$$= {}^5C_3 \times \left(\frac{2}{3}\right)^2 \left(\frac{1}{3}\right)^3$$

$$= 10 \times \frac{4}{243}$$

$$\simeq 0.16$$

Example 7.1

A fair die is tossed eight times, find the probability of obtaining exactly three sixes.

In this example a 'success' will be counted when a six occurs, and since we have a fair die $P(\text{success}) = 1/6$.

Now any sequence containing three 'successes' and five 'failures' will have probability

$$\left(\frac{5}{6}\right)^5 \left(\frac{1}{6}\right)^3 \simeq 0.00186$$

and there are ${}^8C_3 = 56$ such sequences

So

$$P(\text{exactly three successes}) = {}^8C_3 \left(\frac{5}{6}\right)^5 \left(\frac{1}{6}\right)^3$$

$$= 56 \times 0.00186$$

$$\simeq 0.104$$

The Binomial variate

Suppose n independent trials of an experiment are to be performed, where the probability of success is p and the probability of failure is q. (Clearly $q = 1 - p$.)

Let X be the variate which takes the integral value r, between 0 and n inclusive, equal to the number of successes obtained.

Now the probability of a particular sequence consisting of r successes and $n - r$ failures is $q^{n-r} p^r$, by independence. Further there are precisely ${}^nC_r$ such mutually exclusive sequences so

$$P(X = r) = p(r) = {}^nC_r \, q^{n-r} \, p^r$$

We refer to such a random variable X as a Binomial variate or say simply that X is $B(n, p)$ where n is the number of trials and p is the probability of success.

Definition 7.1

If X is a discrete variate which can assume integral values between 0 and n inclusive such that

$P(X = r) = {}^nC_r \, q^{n-r} \, p^r$, where $q = 1 - p$,

then X is a Binomial variate denoted $B(n, p)$

In the case that $n = 4$ we have

$$P(X = 0) = {}^4C_0 \, q^4 p^0 = q^4$$
$$P(X = 1) = {}^4C_1 \, q^3 p = 4q^3 p$$
$$P(X = 2) = {}^4C_2 \, q^2 p^2 = 6q^2 p^2$$
$$P(X = 3) = {}^4C_3 \, qp^3 = 4qp^3$$
$$P(X = 4) = {}^4C_4 \, q^0 p^4 = p^4$$

It is interesting to note that the terms on the right are merely the terms in the expansion of the binomial:

$$(q + p)^4 = q^4 + 4q^3 p + 6q^2 p^2 + 4qp^3 + p^4.$$

This is true in the general case, that is,

$$(q + p)^n = P(0) + P(1) + P(2) + \ldots + P(n).$$

It is this property which determined the name *Binomial distribution*. The distribution was first discovered by an eighteenth century mathematician, James Bernoulli. The trials referred to above are sometimes known as Bernoulli trials – indicating that the trials are independent of one another and result in either a 'success' or a 'failure'.

Example 7.2

A multiple-choice examination consists of 12 questions each of which has five possible answers. A particular candidate decides to select his answers at random. Find the probability that he obtains less than three correct answers.
In this problem we see that there are 12 trials and the probability of a success (obtaining a correct answer by choosing randomly) is $1/5$.
We model this situation by considering the variate X which is $B(12, 1/5)$. So

$$P(X = r) = {}^{12}C_r \left(\frac{4}{5}\right)^{12-r} \left(\frac{1}{5}\right)^r$$

for integral r between 0 and n inclusive. We require

$$P(X < 3) = P(X = 0 \text{ or } X = 1 \text{ or } X = 2)$$

$$= P(0) + P(1) + P(2)$$

$$= {}^{12}C_0 \left(\frac{4}{5}\right)^{12} \left(\frac{1}{5}\right)^0 + {}^{12}C_1 \left(\frac{4}{5}\right)^{11} \frac{1}{5} + {}^{12}C_2 \left(\frac{4}{5}\right)^{10} \left(\frac{1}{5}\right)^2$$

Alternatively we require the sum of the first three terms in the expansion of

$$\left(\frac{4}{5} + \frac{1}{5}\right)^{12} = \left(\frac{4}{5}\right)^{12} + 12 \left(\frac{4}{5}\right)^{11} \frac{1}{5} + \frac{12 \times 11}{2!} \left(\frac{4}{5}\right)^{10} \left(\frac{1}{5}\right)^2 + \ldots,$$

which amounts to the same thing.

In either case

$$P(X < 3) \simeq 0.0687 + 0.2062 + 0.2835$$

$$= 0.558$$

A result which suggests that it is not a good idea to guess the answers in a test!

7.2 The mean and variance of a Binomial distribution

If the variate X is distributed as $B(n, p)$ then clearly n is a positive integer and $0 \leqslant p \leqslant 1$ and we have shown that

$$P(X = r) = P(r) = {}^nC_r\, q^{n-r}\, p^r, \quad \text{where } q = 1 - p.$$

In order to be sure that this is a probability distribution we have to check that

(i) $P(r) > 0$ for $r = 0, 1, 2, \ldots, n$ and this follows from the restrictions on n and p.

(ii) $\Sigma P(r) = 1$

   Now

$$\Sigma P(r) = P(0) + P(1) + P(2) + \ldots + P(n)$$

$$= (q + p)^n, \text{ by the binomial expansion}$$

$$= 1, \text{ from the definition of } q.$$

At this stage we shall state the mean and variance of X. Proofs can be found in p. 150.

Statement 7.1

> If X is $B(n, p)$ then:
>
> $$E(X) = np$$
>
> $$\text{Var}(X) = npq$$

Thus if X is $B(4, \frac{1}{2})$ then

$$E(X) = 4 \times \tfrac{1}{2}$$

$$= 2$$

$$\text{Var}(X) = 4 \times \tfrac{1}{2} \times \tfrac{1}{2} = 1$$

Example 7.3

Suppose four unbiased coins are tossed, find the expected number of 'heads' obtained and the variance.

   Counting a 'success' when a head is obtained we see that $p = \frac{1}{2}$ and clearly $n = 4$. Thus if X represents the number of 'heads' obtained then X is $B(4, \frac{1}{2})$, since the trials are independent.

From the results stated above $E(X) = 2$ and Var $(X) = 1$, however these results can be derived directly as follows.

$$E(X) = 0 \times P(0) + 1 \times P(1) + 2 \times P(2) + 3 \times P(3) + 4 \times P(4)$$

and the probabilities can be found directly from the expansion of

$$(q + p)^n = (\tfrac{1}{2} + \tfrac{1}{2})^4 = \tfrac{1}{2}^4 + 4(\tfrac{1}{2})^3 \tfrac{1}{2} + 6(\tfrac{1}{2})^2 (\tfrac{1}{2})^2 + 4(\tfrac{1}{2})(\tfrac{1}{2})^3 + (\tfrac{1}{2})$$

$$= \frac{1}{16} + \frac{4}{16} + \frac{6}{16} + \frac{4}{16} + \frac{1}{16}$$

Thus

$$E(X) = 0 \times \frac{1}{16} + 1 \times \frac{4}{16} + 2 \times \frac{6}{16} + 3 \times \frac{4}{16} + 4 \times \frac{1}{16}$$

$$= \frac{32}{16}$$

$$= 2$$

Similarly

$$E(X^2) = 0^2 \times \frac{1}{16} + 1^2 \times \frac{4}{16} + 2^2 \times \frac{6}{16} + 3^2 \times \frac{4}{16} + 4^2 \times \frac{1}{16}$$

$$= \frac{80}{16}$$

$$= 5$$

So

$$\text{Var}(X) = E(X^2) - E(X)^2$$

$$= 5 - 2^2$$

$$= 1.$$

Example 7.4
State the conditions under which the Binomial distribution $B(n, p)$ applies.

If an experiment has been designed to result in either a 'success' or a 'failure' with constant probabilities p and q respectively, and n independent trials of the experiment are performed, then the variate X representing the number of successes is $B(n, p)$. Thus

$$P(X = r) = {}^nC_r \, q^{n-r} p^r, \quad \text{for } r = 0, 1, 2, \ldots, n, \text{ where } q = 1 - p$$

Example 7.5
In a certain inspection scheme a sample of ten items is selected at random from a very large batch and the number of defectives is recorded. If this number is more than two the batch is rejected, and it is accepted if there are no defectives. Otherwise a further sample of five items is selected, again at random. If there are any defectives this time, the batch is rejected, otherwise it is accepted.

If the proportion of defectives is in fact 10%, find the probability that

(a) the batch is accepted as a result of the first inspection,
(b) that a further sample is taken,
(c) that the batch is accepted given that a second sample is taken,
(d) the batch is accepted after a second sample is taken,
(e) the batch is accepted.

Let X represent the number of defectives obtained in the first sample of ten items and Y represent the number of defectives obtained in the second sample when taken.

We argue that X is approximately $B(10, 1/10)$ and Y is approximately $B(5, 1/10)$. We know that X is not exactly $B(10, 1/10)$ since the probability of obtaining a defective does not remain constant from one trial to the next, since the number of defectives left depend on how many have previously been selected. We are told, however, that a very large batch of items is available and since we are taking small samples the actual proportion will not vary much from trial to trial so we shall accept that the approximations are satisfactory.

(a) P(batch accepted after first inspection)

$= P(X = 0)$

$= 0.349$

(b) P(further sample taken)

$= P(X = 1 \text{ or } X = 2)$

$= 0.581$

(c) P(batch accepted given that a further sample has already been taken)

$= P(Y = 0)$

$= 0.590$

(d) P(batch will be accepted only after a further sample has been taken)

$P((X = 1 \text{ or } 2) \text{ and } Y = 0)$

$= P(X = 1 \text{ or } 2) \times P(Y = 0)$, by independence

$= 0.581 \times 0.590$

$= 0.343$

(e) P(batch accepted)

$= P(X = 0 \text{ or } (X = 1 \text{ or } 2 \text{ and } Y = 0))$

$= P(X = 0) + P(X = 1 \text{ or } 2 \text{ and } Y = 0)$

$= 0.349 + 0.343$

$= 0.692$

Exercise 7.1

1. Check the results in Example 6.2 above by tossing five coins of your choice and recording the number of heads obtained, to form a frequency distribution.

2. A fair die is tossed eight times. Find the probability of obtaining
   (a) no sixes,
   (b) just one six,
   (c) more than one six.

3. Find the probability that in a family of four children two are boys and two are girls. State any assumption you have made.

4. If X is $B(6, \frac{1}{3})$ find (a) $P(X = 2)$, (b) $P(X \geqslant 5)$, (c) $P(2 < X < 5)$.

5. In a large batch of dresses 10% are faulty. If six are chosen at random which is more likely
   (a) they are all sound,
   (b) at least one of the dresses is faulty?

6. (i) Two dice are loaded so that in each case $P($obtaining 6$) = 1/3$.
      Find the probability of obtaining (a) just one six, (b) two sixes when the two dice are tossed.
   (ii) If the two dice are thrown five times, find the probability of obtaining exactly four sixes.

7. A test is applied to a certain type of component. If the probability that the component fails is 1/5 calculate the following probabilities for a test of 10 components
   (i) that none will fail,
   (ii) that less than two will fail,
   (iii) that exactly two will fail,
   (iv) that more than two will fail.

8. In a cricket match it is claimed that the probability a certain bowler takes a wicket with any ball is 1/20. Calculate the probability of his taking
   (a) three wickets in four balls,
   (b) at least one wicket in six balls.

9. A bag contains five red and three blue balls. Balls are taken out at random and the colour noted. Find the probability of obtaining
   (i) three blue,
   (ii) just one red when four such choices are made,

   (a) assuming the balls are replaced after each choice,
   (b) assuming the balls are not replaced.

10. How many times must an unbiased coin be tossed so that the probability that at least one head is obtained is approximately 0.95?

11. If X is $B(n, p)$ where the mean of X is 10 and the variance 6.67, find the value of n and of p.

12. A man has a probability of $1/6$ of hitting his target with a rifle. How many rounds must be fired in order to have a probability of 0.90 or more of hitting the target at least once?

13. A drunk who is equally likely to take a step forward towards his house as backwards towards the inn, is standing by a lamp-post. After the next ten steps find the probability that he is
    (a) ten steps nearer home,
    (b) back where he started by the lamp-post,
    (c) two steps nearer to the inn.

14. In your own words define the Binomial distribution explaining the meaning of each term introduced and write down the mean and variance of the distribution.

15. Coffee and mint chocolates are mixed in the ratio $4:1$. They are packaged, 16 to a packet, before being sold. Find
    (a) the average number of mint chocolates per packet,
    (b) the probability that there are less than the average number of mint chocolates in a packet chosen at random.

16. Expand $(2/3 + 1/3)^5$ using the binomial expansion, and hence write down the theoretical frequency (to the nearest integer) of obtaining 0, 1, 2, 3, 4, 5 'successes' when five dice are tossed 100 times. A success is counted when either a four or a five is obtained when one die is tossed.

    Use this theoretical frequency table using rounded frequencies to calculate the mean and variance of the number of successes obtained.

    Compare your answers with those given by $100\,np$ and $100\,npq$ (with standard notation).

17. Perform the experiment referred to above by actually tossing 5 dice 100 times. Obtain a frequency table and calculate the mean and variance. Explain, as well as you can, any discrepancy between these results and those obtained in question 16.

18. An experiment is performed consisting of n Bernoulli trials, each having probability of success p. If the experiment is repeated N times show that the theoretical frequency with which r successes are obtained is given by

$$N\,{}^nC_r\,q^{n-r}\,p^r$$

Draw theoretical histograms representing the case

   (a) $N = 100$, $n = 10$, $p = 0.5$;
   (b) $N = 100$, $n = 10$, $p = 0.1$.

19. In a certain gambling game a player nominates an integer x from 1 to 6 inclusive and he then throws three fair cubical dice. Calculate the probabilities that the number of x's thrown will be 0, 1, 2 and 3.

   The player pays 5 pence per play of the game and he receives 48 pence if the number of x's thrown is three, 15 pence if the number of x's thrown is two, 5 pence if only one x is thrown and nothing otherwise. Calculate the player's expected gain or loss per play of the game.

JMB

20. Mass production of miniature hearing aids is a particularly difficult process and so the quality of these products is monitored carefully. Samples of size six are selected regularly and tested for correct operation. The number of defectives in each sample is recorded. During one particular week 140 samples are taken and the distribution of the number of defectives per sample is given in the following table.

Number of defectives per sample (x)	0	1	2	3	4	5	6
Number of samples with x defectives (f)	27	36	39	22	10	4	2

Find the frequencies of the number of defectives per sample given by a Binomial distribution having the same mean and total as the observed distribution.

AEB 1978

7.3 The Geometric distribution

Definition 7.2

> The discrete variate X is said to follow the Geometric distribution with parameter p if X takes the value of any natural number 1, 2, 3, 4, ... such that
>
> $$P(X = r) = (1 - p)^{r-1} p \quad \text{where the constant } p \text{ satisfies } 0 \leqslant p \leqslant 1$$

We can check that the above represents a probability distribution since

(i) $(1 - p)^{r-1} p \geqslant 0$, by the constraint on the value of p.

(ii) Representing $1 - p = q$ the sum of all the probabilities is

$$p + qp + q^2 p + \dots,$$

which is an infinite geometric series with sum

$$\frac{p}{1 - q} = 1$$

as required.

The geometric distribution arises in the following manner. Suppose a Bernoulli trial results in a 'success' with probability p, and X represents the number of trials required before a success occurs.

Then

$P(X = 1) = p,$ since we are dealing with a Bernoulli trial;

$P(X = 2) = qp,$ since one failure is followed by a success;

$P(X = 3) = q^2p,$ since two successive failures are followed by a success.

$$\vdots$$

Thus X follows the geometric distribution.

We state the following facts

Statement 7.2

If X is a variate following the geometric distribution with parameter p then:

$$E(X) = \frac{1}{p} \qquad \text{Var } X = \frac{q}{p^2}$$

A proof can be found on p. 151.

Example 7.6

Two players A and B toss a die in turn until a six appears, then the person who threw it has won. Find the probability that A wins the game and find the average number of tosses required before the game is decided.

$P(A \text{ wins on his first turn}) \qquad = 1/6$

$P(A \text{ wins on his second turn}) \quad = \left(\frac{5}{6}\right)^2 \times \frac{1}{6}$

$P(A \text{ wins on his third turn}) \qquad = \left(\frac{5}{6}\right)^4 \times \frac{1}{6}$

and generally

$P(A \text{ wins on his } r\text{th turn}) \qquad = \left(\frac{5}{6}\right)^{2r-2} \times \frac{1}{6}$

Thus $P(A \text{ wins the game})$

$$= \frac{1}{6} + \left(\frac{5}{6}\right)^2 \frac{1}{6} + \left(\frac{5}{6}\right)^4 \frac{1}{6} + \dots + \left(\frac{5}{6}\right)^{2r-2} \frac{1}{6} + \dots,$$

which is an infinite geometric series with sum

$$\frac{\frac{1}{6}}{1 - \left(\frac{5}{6}\right)^2} = \frac{6}{11},$$

giving A the advantage in this game.

Let Y represent the number of tosses required before the game is decided.

The probability distribution of Y is as follows

Y	1	2	3	4	... r ...
p_r	$\dfrac{1}{6}$	$\left(\dfrac{5}{6}\right)\left(\dfrac{1}{6}\right)$	$\left(\dfrac{5}{6}\right)^2\left(\dfrac{1}{6}\right)$	$\left(\dfrac{5}{6}\right)^3\left(\dfrac{1}{6}\right)$	... $\left(\dfrac{5}{6}\right)^{r-1}\dfrac{1}{6}$ ...

Now $E(Y) = \Sigma r \left(\dfrac{5}{6}\right)^{r-1}\dfrac{1}{6}$, where the summation takes r from 1 to infinity,

$$= \frac{1}{6\left(1 - \dfrac{5}{6}\right)^2},$$ to see this, simply expand $\dfrac{1}{6}\left(1 - \dfrac{5}{6}\right)^{-2}$

using the Binomial expansion

$$= 6$$

Alternatively, noting that Y follows the geometric distribution with parameter 1/6, then

$$E(Y) = \frac{1}{\dfrac{1}{6}}$$

$$= 6$$

So on average six tosses are required before the game is decided.

Exercise 7.2

1. A coin is tossed until the first head appears. It is found on average that three tosses are required. Find the probability of obtaining a head with this coin.

2. If the variate X follows the geometric distribution with parameter p show that $E(X) = 1/p$ directly, using the fact that

$$(1 - x)^{-2} = 1 + 2x + 3x^2 + 4x^3 + ...,$$

whenever $|x| < 1$.

3. Two players A and B, toss a fair coin. A, who starts the game, stakes a penny each time he throws the coin; B also stakes a penny at each of his throws. The first player to throw heads wins and gathers all the stakes. Find the probability that A wins the game. Explain why A, although he wins more frequently, loses money. Calculate his expected loss on 100 games.

[Hint: for $|z| < 1$, $1 + 2z + 3z^2 + 4z^3 + ... = (1 - z)^{-2}$.]

MEI

4. (*a*) A fair six sided die is thrown and if a six is obtained the die is thrown again, repeating until a number other than six is obtained. Find the probability distribution of the number of throws. Find the probability that the sum of the numbers thrown is (i) less than 3, (ii) exactly 15, (iii) exactly 18.

(*b*) A fair six-sided die is thrown, repeatedly if necessary, until the first six is obtained. Find (i) the probability that the number of throws is greater than two, (ii) the mean of the number of throws.

$$[1 + 2x + 3x^2 + \ldots = (1 - x)^{-2}]$$

JMB

5. At a shooting gallery, each time a marksman fires the probability of his hitting a clay pipe is 1/4.

What is the probability of his first hitting a pipe
(*a*) in four shots or less,
(*b*) with his third shot?

The marksman pays 3p for each shot and gets a reward of 15p when he hits a pipe.

Show that he may expect to win 3p if he plays the game until he hits a pipe.

$$[\text{You may assume that } 1 + 2x + 3x^2 + \ldots = (1 - x)^{-2}.]$$

LOND

Revision exercise A

1. The distribution of ages of 90 people borrowing books at a particular library is given as follows.

Age in completed years	8–12	13–20	21–60	61–64
Number of people	10	24	40	16

Represent this data in the form of a histogram. Estimate the mean and standard deviation of the ages of the 90 people giving your answer correct to 1 DP.

2. 121 students recorded the number of half-days absence during a particular school term. The data is recorded below.

Number of half-days absence	0	1–10	11–30	31–60
Number of students	30	65	24	2

Represent this data in the form of a histogram and estimate the mean and variance.

3. Below are given the number n of hours worked in a week by 64 men.

```
36.5  15.6  30.8  27.6  33.6  39.4  39.7  39.1
33.0  40.1  21.8  40.6  33.9  36.9  39.1  27.4
29.8  37.0  45.4  42.5   9.6  26.3  36.1  35.2
38.4  28.5  30.5  44.4  38.4  40.6  26.5  38.7
24.4  41.9  52.7  35.7  28.9  38.2  30.4  41.7
43.2  24.5  34.8  37.8  38.0  43.7  40.8  11.7
31.5  37.6  40.1  23.7  31.8  42.0  29.1  41.5
36.3  29.7  37.3  28.4  39.6  22.9  35.2  42.4
```

(i) Group the numbers into intervals of width 3 hours defined by $9.5 \leqslant n < 12.5$, $12.5 \leqslant n < 15.5, \ldots$

(ii) Use the grouped data to calculate estimates of the mean and standard deviation of n.

(iii) Estimate the percentage of workmen for whom n is within one standard deviation of the mean.

MEI

4. (a) State the addition and multiplication rules of probability, explain what is meant by the terms and show that:
(i) mutually exclusive $\Leftrightarrow P(A|B) = 0$
(ii) Statistical independence $\Leftrightarrow P(A|B) = P(A)$

(b) What is the probability of drawing two black cards from a pack of cards if
(i) the first card is replaced,
(ii) the first card is not replaced?

5. A builder has to choose between two jobs. The first will make a profit of £16 000 with probability 0.75 or a loss of £8000. The other job has a probability of 1/3 of making a profit of £36 000 otherwise a loss of £6000. Which job should he choose?

6. Four counterfeit £1 notes are mixed with 16 genuine £1 notes. If two notes are selected at random find the probability that
(a) neither note is counterfeit,
(b) both notes are counterfeit.

7. The following are three of the classical problems in probability.
(a) Compare the probability of a total of 9 with the probability of a total of 10 when three fair dice are tossed once (Galileo and Duke of Tuscany).
(b) Compare the probability of at least one six in 4 tosses of a fair die with the probability of at least one double-six in 24 tosses of two fair dice (Chevalier de Mere).
(c) Compare the probability of at least one six when 6 dice are rolled with the probability of at least two sixes when 12 dice are rolled (Pepys to Newton).

Solve each of these problems.

AEB 1978

8. A box contains 9 discs, of which 4 are red, 3 are white and 2 are blue. Three discs are to be drawn at random without replacement from the box. Calculate
   (a) the probability that the discs, in the order drawn will be coloured red, white and blue, respectively,
   (b) the probability that one disc of each colour will be drawn,
   (c) the probability that the third disc drawn will be red,
   (d) the probability that no red disc will be drawn,
   (e) the most probable number of red discs that will be drawn,
   (f) the expected number of red discs that will be drawn, and state the probability that this expected number of red discs will be drawn.

   JMB

9. A faulty gas-lighter works successfully, on average, once in three attempts. When using it one goes on trying until it does work and the gas lights.
   What is the probability that the gas lights (a) first time, (b) second time, (c) at the fifth attempt?
   What is the probability that it will not have worked after five attempts?
   The sellers of the lighter promise that if one is returned, they will pay 25p for each time it fails to work before the gas is lit. What is one's expectation in taking up this offer? What is the probability of getting one's money back (or more) if the original price was £1?
   If four people take up this offer, what is the probability that at least two get their money back (or more)?

   AEB 1978

10. (a) A gambler pays £1 to draw six cards, one from each of six ordinary packs. If at least four of the cards are spades, he receives £26. Otherwise he receives nothing. How much can he expect to lose in 100 tries?
    (b) The game is now changed, so that the gambler draws six cards from one pack, without replacement, and receives £30 if there are four or more spades among the six drawn.
       What is a fair price for him to pay?

    AEB 1976

11. (a) A man throws two fair dice and receives a number of pence equal to the product of the numbers of spots shown by the uppermost faces of the dice.
       What is his expectation at each throw of the dice?
    (b) An electric circuit contains 20 components, each of which has a probability 0.01 of being defective.
       Calculate the probability that none of the components is defective.
    (c) A bag contains 3 red counters, 4 white counters and 5 blue counters. Two counters are drawn at random. Find the probability that they are of different colours.

    AEB 1976

12. A lake contains 1000 fish of species A and 2000 fish of species B and no other fish. Ten fish are taken from the lake at random. Explain why the probabilities of obtaining various numbers of fish of species A in the ten can be very well approximated by a Binomial probability distribution, and hence estimate to three significant figures the probability of obtaining 4 fish of species A and 6 of species B.

Estimate also to three significant figures the probability that in a catch of 5 fish altogether there are more fish of species A than there are of species B.

CAMB

13. A random sample of 1000 surnames is drawn from a local telephone directory. The distribution of the lengths of the names is as shown.

Number of letters in surname	3	4	5	6	7	8	9	10	11	12
Frequency	13	102	186	237	215	113	83	32	13	6

Calculate the sample mean and sample standard deviation. Obtain the upper quartile.

Represent graphically the data in the table.

Give a reason why the sample of names obtained in this way may not be truly representative of the population of Great Britain.

JMB

14. Easyadd Ltd. manufacture electronic desk calculators. One of the electronic components in the calculator is available in two types, A and B. In order to decide which to use in their calculator they obtain a sample of 100 of each type and test these to failure. The results of these tests are summarized below:

Time to failure (hours)	Number of components	
	Type A	Type B
Less than 10	10	1
10–19	10	2
20–29	9	3
30–49	9	9
50–99	7	23
100–149	8	18
150–199	6	11
200–299	8	13
300–399	5	8
400–499	3	5
500 and over	25	7
Total	100	100

Compare these two distributions (i) graphically, (ii) numerically.

What is your interpretation of the differences, and which type of component would you recommend the company to use?

AEB 1975

Chapter 8   Probability density functions

8.1 Introduction to probability density functions

In previous chapters each variate we have met has been a *discrete* variable; its value has been obtained by *counting*. For instance, the score obtained when tossing a die or the number of heads obtained when five coins are thrown. It is possible to describe the probability for which each of these values occur by means of a probability distribution (cf. 6.2). In essence, this is a function from the set of values the variate can assume to the real numbers between 0 and 1, representing probabilities, such that the sum of these probabilities is 1.

Suppose, however, we wished to let X represent the height of a person, the length of a rod or the temperature of a liquid. In each case X is a *continuous* variate, and the value of X is obtained by *measuring*. In this case, as we shall see, the probability that X assumes any particular value is zero!

For example, suppose that a real number X is chosen at random between 0 and 2.

We would expect

$$P(0 \leqslant X \leqslant 1) = \frac{1}{2}$$

$$P(a \leqslant X \leqslant b) = \frac{b - a}{2} \qquad \text{if } 0 \leqslant a < b \leqslant 2$$

$$P(t \leqslant X \leqslant t + \delta t) = \frac{\delta t}{2} \qquad \text{if } 0 \leqslant t < 2$$

and thus in the limit as $\delta t \to 0$

$$P(t \leqslant X \leqslant t) = P(X = t) = 0 \qquad\qquad \text{!!}$$

Again, suppose Y is a variable representing the length of blades of a particular variety of grass after one year's growth under controlled conditions. If a sample is obtained and the length of each blade is established to the nearest cm, the following histogram might result. (Fig. 8.1(a)).

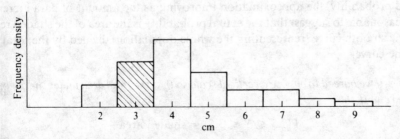

Fig. 8.1(a)   Histogram to show the lengths of a sample of blades of grass of a particular variety measured to the nearest cm.

If a larger sample were obtained it might be worthwhile to measure to the nearest mm, with the resulting histogram. (Fig. 8.1(*b*)).

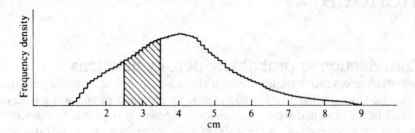

Fig. 8.1(*b*) Histogram to show the length of a sample of blades of grass of a particular variety measured to the nearest mm.

As we increase the amount of data we could, if desired, measure the blades even more accurately. Further, if the blades of grass were chosen randomly then the resulting histograms would be more and more accurate representations of the whole population of lengths of blades of this particular variety of grass. We are tempted to use a smooth curve to represent this distribution. Perhaps the following would be suitable. (Fig. 8.1 (*c*))

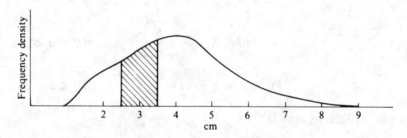

Fig. 8.1(*c*) The limiting form of a histogram.

Suppose we were concerned to find the probability that a blade of grass chosen at random had a length between 2.5 and 3.5 cm. In each of the above histograms we would accept that the shaded area divided by the total area was an approximation to the required probability, the approximation improving as the amount of data increased.

It is reasonable to suggest that the actual probability is the area of the shaded region under the smooth curve (representing the whole population) divided by the total area under the curve.

Further, if we agreed to choose a vertical scale so that the total area under the curve was one square unit then

$$P(2.5 \leqslant Y \leqslant 3.5) = \text{shaded area}$$

$$= \int_{2.5}^{3.5} \left(\begin{array}{l} \text{the function of } Y \\ \text{represented by the curve} \end{array} \right) dy$$

and more generally, if the function represented by the curve was $f(y)$ then

$$P(a \leqslant Y \leqslant b) = \int_a^b f(y)\,dy$$

This leads us to make the following definition.

Definition 8.1

> If X is a continuous variable and $f(x)$ is a function for which
>
> (i) $f(x) \geqslant 0$ for all x
>
> (ii) $\displaystyle\int_{-\infty}^{\infty} f(x)\,dx = 1$
>
> then $f(x)$ is said to be a *probability density function* for X.

Note our continued use of the (lower-case) letter x to represent the values the (upper case) variable X can assume.

Thus probability density functions (pdf for short) are for the continuous variable what probability distributions are for the discrete variable. In this case, however, integration can be used to establish probabilities. We shall continue to use the term 'random variable' or 'variate' when talking of a continuous variable with an associated pdf.

Example 8.1
The length X (in cm) of blades of another variety of grass after one year's growth is thought to be distributed as indicated in the diagram and have a pdf given by

$$f(x) = \begin{cases} Cx(4 - x) & 0 \leqslant x \leqslant 4 \\ 0 & \text{otherwise} \end{cases}$$

where C is a positive constant. (It is a common practice to use 'part' of a function to form a pdf.) Find the value of C and also the probability that a blade of grass chosen at random will be less than 1 cm in length.

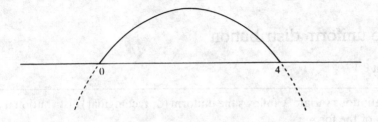

To find the value of the constant C use the fundamental property of pdf's.

$$\int_{-\infty}^{\infty} f(x)dx = 1, \quad \text{which clearly reduces to}$$

$$\int_{0}^{4} Cx(4 - x)dx = 1, \quad \begin{array}{l}\text{since elsewhere the area} \\ \text{under the curve is zero}\end{array}$$

$$\Rightarrow C \int_{0}^{4} 4x - x^2 \, dx = 1$$

$$\Rightarrow C \left[2x^2 - \frac{x^3}{3} \right]_{0}^{4} = 1$$

$$\Rightarrow C \frac{32}{3} = 1$$

$$\Rightarrow C = \frac{3}{32}.$$

$$P(0 \leqslant X \leqslant 1) = \frac{3}{32} \int_{0}^{1} x(4 - x)dx, \quad \text{now that } C \text{ is known.}$$

$$= \frac{3}{32} \left[2x^2 - \frac{x^3}{3} \right]_{0}^{1}$$

$$= \frac{5}{32}.$$

We can imagine the process by which the particular pdf for X was established.
Suppose a large random sample of blades of grass was obtained, the length measured and a histogram drawn to represent the data. The symmetry and general shape might suggest that we use a quadratic to represent the pdf. The particular member of this family of possible quadratics is then established by using the fundamental property of a pdf.

We could even test our 'model' by calculating various probabilities of interest and comparing these answers with those obtained from the sample.

8.2 The uniform distribution

Statement 8.1

A continuous variate X follows the uniform (or rectangular) distribution if its pdf is of the form

$$f(x) = \begin{cases} \dfrac{1}{b - a} & a \leqslant x \leqslant b \\ 0 & \text{otherwise} \end{cases}$$

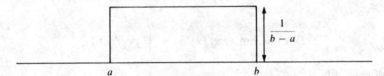

Fig. 8.2   The uniform distribution.

It is easy to check that this is a valid pdf by integration, though in this case the area of the rectangle is readily seen to be one unit.

In the next exercise the reader will be asked to establish the following results.

Statement 8.2

If X is a continuous variate with a Uniform distribution between the limits a and b, where $a < b$ then

$$E(X) = \frac{a+b}{2}$$

$$\text{Var}(X) = \frac{(b-a)^2}{12}$$

8.3 Expectation and variance

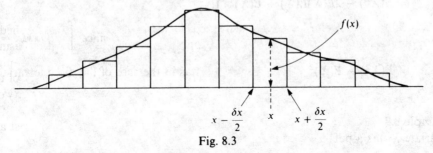

Fig. 8.3

As previously noted, for a continuous variate X, the probability that X equals any particular value is zero. The probability that X is 'close' to the value x will mean

$$P(X \simeq x) = P\left(x - \frac{\delta x}{2} \leqslant X \leqslant x + \frac{\delta x}{2}\right)$$

and as the diagram Fig. 8.3 suggests

$$P(X \simeq x) \simeq f(x)\delta x, \quad \text{where } f(x) \text{ is the pdf of } X.$$

If we approximate the pdf $f(x)$ by a histogram with class intervals of width δx and mid interval values $x_1, x_2, x_3, \ldots x_n$ then we would argue that the expectation of these values was

$$\Sigma x P(X \simeq x) \simeq \Sigma x f(x)\delta x$$

In the limit as $\delta x \to 0$ it seems reasonable to define, in the case of the continuous variable X,

$$E(X) = \lim_{\delta x \to 0} (\Sigma x f(x)\delta x) = \int_{-\infty}^{\infty} x f(x)dx,$$

and more generally

$$E(g(X)) = \int_{-\infty}^{\infty} g(x)f(x)dx, \qquad (*)$$

where $g(X)$ is a function of X and $f(x)$ is the pdf of X.

The formal definition of the variance of X is

$$\text{Var}(X) = E([X - E(X)]^2)$$

$$= \int_{-\infty}^{\infty} [x - E(X)]^2 f(x)dx, \qquad \text{by}(*)$$

$$= \int_{-\infty}^{\infty} [x^2 - 2x\,E(X) + E(X)^2]f(x)dx$$

$$= \int_{-\infty}^{\infty} x^2 f(x)dx - 2E(X)\int_{-\infty}^{\infty} xf(x)dx + E(X)^2 \int_{-\infty}^{\infty} f(x)dx,$$

since $E(X)$ is a fixed value and so can be taken outside the integral.

$$= E(X^2) - 2E(X)E(X) + E(X)^2 . 1,$$

since $\displaystyle\int_{-\infty}^{\infty} f(x)dx = 1$

$$= E(X^2) - E(X)^2$$

as in the case of the discrete variate.

Example 8.2
A variate X has a pdf

$$f(x) = \begin{cases} kx(2 - x) & 0 \leqslant x \leqslant 2 \\ 0 & \text{otherwise} \end{cases}$$

Find the value of the constant k and the mean and variance of X. Find also the probability that two values of X chosen at random from such a distribution will both be larger than 1.

To find k we use the basic property of pdf's

$$\int_0^2 kx(2 - x)dx = 1$$

$$\Rightarrow k\int_0^2 2x - x^2\, dx = 1$$

$$\Rightarrow k \left[x^2 - \frac{x^3}{3} \right]_0^2 = 1$$

$$\Rightarrow k \frac{4}{3} = 1$$

$$\Rightarrow k = \frac{3}{4}$$

The pdf is symmetrical about the point $X = 1$, as a sketch will show and thus the mean is 1. More formally we have

$$E(X) = \int_0^2 x \cdot \frac{3}{4} x(2 - x) dx$$

$$= \frac{3}{4} \int_0^2 2x^2 - x^3 \, dx$$

$$= \frac{3}{4} \left[\frac{2x^3}{3} - \frac{x^4}{4} \right]_0^2$$

$$= \frac{3}{4} [5\tfrac{1}{3} - 4]$$

$$= 1.$$

To obtain the variance we could use

$$\text{Var}(X) = E[(X - E(X))^2]$$
$$= E[(X - 1)^2], \text{ since } E(X) = 1$$
$$= \int_0^2 (x - 1)^2 \frac{3}{4} x(2 - x) \, dx$$

and continue by integration.

Alternatively

$$\text{Var}(X) = E(X^2) - E(X)^2$$

$$E(X^2) = \frac{3}{4} \int_0^2 x^3 (2 - x) \, dx$$

$$= \frac{3}{4} \left[\frac{x^4}{2} - \frac{x^5}{5} \right]_0^2$$

$$= \frac{6}{5}.$$

So

$$\text{Var}(X) = \frac{6}{5} - 1^2 = \frac{1}{5}.$$

The probability that a value of x chosen at random is larger than 1 can be found by integration as

$$P(1 \leqslant x \leqslant 2) = \frac{3}{4} \int_1^2 x(2 - x)\,dx,$$

$$= \frac{1}{2}.$$

and hence

P(both values of X chosen at random, and hence independently, lie between 1 and 2)

$$= \frac{1}{2} \times \frac{1}{2}$$

$$= \frac{1}{4}$$

8.4 The mode and median for a continuous variate

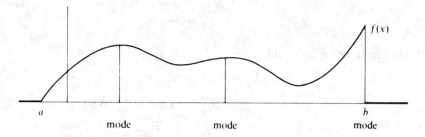

Fig. 8.4   A diagram indicating the position of the modes in a particular pdf $f(x)$.

Suppose the diagram in Fig. 8.4 represents the graph of the pdf $f(x)$ of X which takes values between a and b. The mode (or modes) is (are) the point(s) corresponding to a (local) maximum and apart from the end points can often be found by the use of calculus. This is the equivalent of the definition given in 2.1 for the discrete variate.

The median can also be found by analogy with the discrete case as the point which cuts the area in half, so that P (a value of X chosen at random is less than the median) = P (a value of X chosen at random is greater than the median) = $1/2$. Thus

$$\int_a^M f(x)\,dx = \frac{1}{2}, \quad \text{where } M \text{ is the median.}$$

Naturally, when the pdf is symmetrical about some particular value M of X then this will be the mean as well as the median.

It is worth noting that the mean corresponds to the 'centre of gravity'. That is, if we imagine the area between the pdf and the X axis cut out of cardboard then the point along the X axis about which the figure will balance will be found to be the mean.

Example 8.3

The variate X is claimed to have a pdf

$$f(x) = \begin{cases} 4(x - x^3) & 0 \leqslant x \leqslant 1 \\ 0 & \text{otherwise} \end{cases}$$

Show that $f(x)$ could indeed represent a pdf and, assuming the claim is correct, find the mode and median of X.

To show that $f(x)$ could represent a pdf we need to show that

(i) $$\int_0^1 4(x - x^3)\,dx = 1,$$

(ii) $$f(x) \geqslant 0 \quad \text{for all } x$$

that is $4(x - x^3) \geqslant 0$ for $0 \leqslant x \leqslant 1$

(i) $\displaystyle\int_0^1 4(x - x^3)\,dx$

$= [2x^2 - x^4]_0^1$

$= 1$

(ii) To show that $4(x - x^3) \geqslant 0$ for $0 \leqslant x \leqslant 1$ it suffices to draw a sketch graph. Alternatively we solve

$$4(x - x^3) = 0$$

$$\Rightarrow x(1 - x^2) = 0$$

$$\Rightarrow x(1 - x)(1 + x) = 0$$

$$\Rightarrow x = 0 \quad \text{or} \quad x = 1 \quad \text{or} \quad x = -1$$

Thus the pdf is zero at $x = 0$ and $x = 1$ but not between, and since the integral is 1 this implies that the function is positive between 0 and 1.

To find the mode, use calculus to find any maxima.

$$f = 4(x - x^3)$$

$$\frac{df}{dx} = 4 - 12x^2$$

$$\frac{d^2f}{dx^2} = -24x$$

For a turning value $\dfrac{df}{dx} = 0$

$$0 = 4 - 12x^2$$

$$x = \frac{1}{\sqrt{3}} \quad \text{or} \quad x = \frac{-1}{\sqrt{3}}$$

Since $x = \dfrac{-1}{\sqrt{3}}$ has no interest for us we consider

$$\frac{d^2f}{dx^2} \quad \text{when} \quad x = \frac{1}{\sqrt{3}}$$

and note that

$$\frac{d^2f}{dx^2} < 0 \Rightarrow \text{a maximum}$$

Thus there is a mode at $x = \dfrac{1}{\sqrt{3}} \simeq 0.58$

To find the median, M, note that

$$\int_0^M 4(x - x^3)\,dx = \frac{1}{2}$$

$$\Rightarrow [2x^2 - x^4]_0^M = \frac{1}{2}$$

$$\Rightarrow 2M^2 - M^4 = \frac{1}{2}$$

Such equations can be difficult to solve, but in this case we have

$$2M^4 - 4M^2 + 1 = 0$$

which is a quadratic in M^2. Thus

$$M^2 = \frac{4 \pm \sqrt{8}}{4}$$

$$\Rightarrow M^2 = 1.707 \quad \text{or} \quad M^2 = 0.293$$

$$\Rightarrow M = \pm 1.307 \quad \text{or} \quad M = \pm 0.541$$

The only solution of any value to us is

$$M = 0.541.$$

Exercise 8.1 (Harder examples will be found in Exercise 19.1

1. A variate X can assume values between 0 and 9 with pdf

$$f(x) = \begin{cases} kx(9 - x) & 0 \leqslant x \leqslant 9 \\ 0 & \text{otherwise} \end{cases}.$$

Find the value of k and show that

$$P(0 \leqslant x \leqslant 3) = \frac{7}{27}$$

2. A variate X has a pdf

$$f(x) = \begin{cases} \dfrac{x}{8} & \text{for } 0 \leqslant x \leqslant 4 \\ 0 & \text{otherwise} \end{cases}$$

Sketch the pdf of X and calculate the mean and variance of X.

   If two values are chosen independently and at random find the probability that they both exceed 1.

3. A variate X can assume any value between 2 and 4 and the equation of its pdf is

$$f(x) = \begin{cases} k & 2 \leqslant x \leqslant 4 \\ 0 & \text{otherwise} \end{cases}.$$

Sketch the curve and state the value of k. Find the mean and variance.

4. The continuous variate X has a uniform distribution over its range 0 to 10. Sketch the distribution and write down the mean and median. Calculate the variance of X.

5. Repeat question 4 if the range of values X can assume is between

   (i) 0 and a $a > 0$
   (ii) a and b $b > a$

6. Find the mode of X where X has the pdf

$$f(x) = \begin{cases} k(9 + 12x - x^3) & \text{for } 0 \leqslant x \leqslant 3 \\ 0 & \text{otherwise} \end{cases}.$$

7. A silicon chip has a life of X hours. If X has a pdf $200/x^2$ where $x \geqslant 200$ find the probability that neither of two such chips chosen at random will have to be replaced during the first 300 hours of operation.

8. The probability that a baker will have sold all of his loaves X hours after baking is given by the pdf

$$f(x) = \begin{cases} k(36 - x^2) & \text{for } 0 \leqslant x \leqslant 6 \\ 0 & \text{otherwise} \end{cases}.$$

Determine the value of k and sketch the pdf. Calculate the mean value and the probability that the baker will have some bread left after five hours.

9. A random variable has a pdf given by

$$f(x) = \begin{cases} \dfrac{k}{x^3} & 1 \leqslant x \leqslant 2 \\ 0 & \text{otherwise} \end{cases}.$$

Find

(a) $P(X \leqslant 1.5)$,
(b) the value of y such that $P(X \geqslant y) = 2P(X \leqslant y)$,
(c) the median.

10. A variate X has a pdf given by

$$f(x) = \begin{cases} \dfrac{4}{3x^2} & 1 \leqslant x \leqslant a \\ 0 & \text{otherwise} \end{cases}$$

(i) Find the value of a.
(ii) Find y so that $P(1 \leqslant X \leqslant y) = 1/4$.

11. The pdf of the variate X is given by the function

$$h(x) = \begin{cases} A(12 + 4x - x^2) & 0 \leqslant x \leqslant 6 \\ 0 & \text{otherwise} \end{cases}$$

Find the mean value of X and the probability that a value of X chosen randomly is greater than 4.

12. A random variable X has a pdf

$$f(x) = \begin{cases} C & \text{if} \quad -4 \leqslant x \leqslant -2, \\ & \text{or} \quad 2 \leqslant x \leqslant 4 \\ 0 & \text{otherwise.} \end{cases}$$

(a) Sketch the distribution
(b) Find the value of C and the standard deviation of X
(c) Find the probability of obtaining a value of X within one standard deviation of the mean.

13. A variate X has a pdf given as follows

$$g(x) = \begin{cases} \dfrac{1}{4} & \text{for} \quad 0 \leqslant x \leqslant 2 \\ \dfrac{k}{8} & \text{for} \quad 4 \leqslant x \leqslant 8. \\ 0 & \text{otherwise} \end{cases}$$

Sketch the pdf and hence find the value of k.
Find
(i) the mean,
(ii) the variance,
(iii) $P(1 \leqslant x \leqslant 6)$.

14. Rulers of nominal length 1 m are measured accurately and found to be within 1 mm of the nominal length. We approximate to the actual distribution by using a pdf of the form indicated below, where X represents the error in mm.

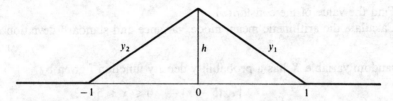

(a) Find the value of h and hence the equations of the lines y_1 and y_2.
(b) Find the standard deviation of the error X.
(c) Find the probability that a ruler picked at random is inaccurate by more than 0.5 mm.

15. The pdf of X is given by

$$f(x) = \begin{cases} \dfrac{x}{24} & \text{for} \quad 0 \leqslant x \leqslant 4 \\ \dfrac{1}{4} - \dfrac{x}{48} & \text{for} \quad 4 < x \leqslant 12 \\ 0 & \text{for} \quad x < 0 \quad \text{or} \quad x > 12 \end{cases}$$

Sketch the pdf of X and find $P(2 < x < 6)$. Write down the mode of X and calculate the median and the mean.

16. The length X of an offcut of wooden planking is a random variable which can take any value up to 0.5 m. It is known that the probability of the length being not more than x m $(0 \leqslant x \leqslant 0.5)$ is equal to kx. Determine
(a) the value of k,
(b) the probability density function of X,
(c) the expected value of X,
(d) the standard deviation of X (correct to 3 significant figures).

17. Petrol is delivered to a garage every Monday morning. At this garage the weekly demand for petrol, in thousands of units, is a continuous random variable X distributed with a probability density function of the form

$$f(x) = \begin{cases} ax(b - x) & 0 \leqslant x \leqslant 1 \\ 0 & \text{otherwise} \end{cases}.$$

(i) Given that the mean weekly demand is 600 units, determine the values of a and b.
(ii) If the storage tanks at this garage are filled to their total capacity of 900 units every Monday morning, what is the probability that in any given week the garage will be unable to meet the demand for petrol?

AEB 1975

18. A random variable x has a probability density function

$$f(x) = Ax(6 - x)^2, \quad 0 \leqslant x \leqslant 6$$

$$= 0 \quad \text{elsewhere}$$

Find the value of the constant A.

Calculate the arithmetic mean, mode, variance and standard deviation of x.

AEB 1975

19. A random variable X has a probability density function f given by

$$f(x) = \begin{cases} cx(5 - x), & 0 \leqslant x \leqslant 5, \\ 0, & \text{otherwise.} \end{cases}$$

Show that $c = 6/125$ and find the mean of X.

The lifetime X in years of an electric light bulb has this distribution. Given that a lamp standard is fitted with two such new bulbs and that their failures are independent, find the probability that neither bulb fails in the first year and the probability that exactly one bulb fails within two years.

MEI

Chapter 9 The Normal distribution

9.1 Introduction

In the previous chapter the notion of a probability density function was introduced for a continuous variate. It was noted that such theoretical distributions could be used to represent or model distributions encountered in the real world.

Without doubt the most important theoretical distribution for a continuous variate is the one called the *Normal distribution*. The diagram in Fig. 9.1 shows the typical shape of the normal distribution.

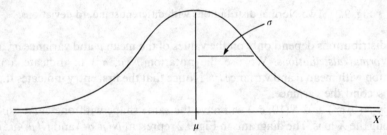

Fig. 9.1 The Normal distribution.

It is symmetrical about the mean and is 'bell' shaped. Most of the distribution is contained within three standard deviations of the mean although it is possible for X to take any real value.

It was first discovered by de Moivre in 1753 as a limiting form of the Binomial distribution. This idea is developed in Section 9.3. Later it was rediscovered by Gauss as the distribution of errors of measurement. In view of this it is not surprising that the curve is often referred to as the Gaussian or 'error' curve. For some time it was thought to be the normal distribution for many natural phenomena. However its main importance is due to its theoretical properties, particularly in sampling theory, which is discussed in Chapter 12.

Statement 9.1

> The pdf of the normal distribution is very complicated. It is in fact
>
> $$\phi(x) = \frac{1}{\sigma\sqrt{2\pi}} e^{\left[-\frac{1}{2}\left(\frac{x-\mu}{\sigma}\right)^2\right]},$$
>
> where e is a constant, μ is the mean and σ the standard deviation of the variate.

In Chapter 19 the reader will be asked to show that μ and σ introduced in the above formula do indeed represent the mean and standard deviation of the distribution.

Different values of μ and σ will give different shapes, but each will remain symmetrical, 'bell' shaped and contain most of the distribution within three standard deviations of the mean.

It is also true that approximately 68% of the distribution is contained within one standard deviation of the mean and 95% is contained within 1.96 standard deviations of the mean.

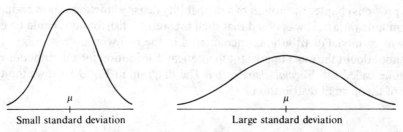

Small standard deviation   Large standard deviation

Fig. 9.2   Two Normal distributions with different standard deviations.

Such distributions depend only on the values of the mean μ and variance σ^2 and are called *Normal distributions*. We use the notation $N(\mu, \sigma^2)$ to indicate a normal distribution with mean μ and variance σ^2. Notice that the first entry indicates the mean and the second the variance.

Clearly $N(0, \sigma^2)$ and $N(10, \sigma^2)$ will have the same shape with one curve shifted 10 units along the X axis. The diagrams in Fig. 9.2 represent $N(\mu, \sigma^2)$ and $N(\mu, 4\sigma^2)$ where changing the variance has had the effect of stretching the curve along the X axis.

Example 9.1

A machine producing rulers of nominal length 30 cm is examined carefully and found to produce rulers whose actual lengths are distributed $N(30, 0.0001)$. Find the probability that a ruler chosen at random has a length between 30 cm and 30.01 cm.

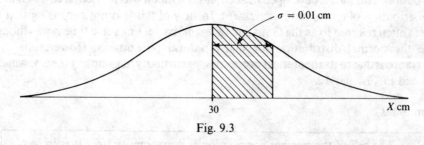

$\sigma = 0.01$ cm

30 X cm

Fig. 9.3

The probability we are trying to find is equal numerically to the area indicated in the diagram in Fig. 9.3. We would expect to find this area by integration, but the pdf, $\phi(x)$, given above is impossible to integrate except by numerical methods.

However, we have stated that approximately 68% of the lengths will be within one standard deviation of the mean, so by symmetry approximately 34% will be in the interval from 30 cm to 30.01 cm. Hence the probability is 0.34.

To find the probability of other events, tables can be used as found on page 303.

These tables are constructed so as to give the probability of being less than Z standard deviations above the mean, such probabilities being denoted by $\Phi(Z)$.

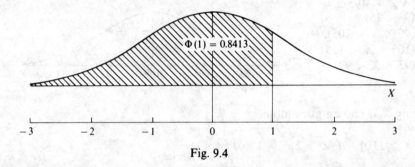

Fig. 9.4

Two points should be noted about such tables. Firstly that no negative values of Z need to be quoted in the tables since (as we shall see) symmetry enables us to calculate the corresponding probabilities. Secondly, it can be shown that only the number of standard deviations from the mean affects the probability, rather than the actual mean or standard deviation.

Example 9.2
A variate X has mean 10 and standard deviation 2.

Find the value of Z, the number of standard deviations from the mean, when X takes the value

(a) 14,   (b) 8,   (c) a.

(a) $Z = \dfrac{14 - 10}{2} = 2$

(b) $Z = \dfrac{8 - 10}{2} = -1$

(c) $Z = \dfrac{a - 10}{2}$

More generally

Statement 9.2

If X is $N(\mu, \sigma^2)$ then the variate

$$Z = \frac{X - \mu}{\sigma},$$

the number of standard deviations X is from the mean, is called the standardised variate and is $N(0, 1)$.

Example 9.3

A variate X is $N(100, 25)$, find the following probabilities

(a) $P(X \geqslant 110)$,

(b) $P(X < 95)$,

(c) $P(95 < X < 105)$,

(d) $P(105 < X < 115)$,

(e) $P(93 < X < 99)$,

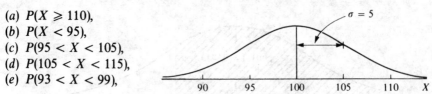

assuming X is chosen at random.

(a) $P(X \geqslant 110) = P(Z \geqslant 2)$ $= 1 - P(Z \leqslant 2)$

$$= 1 - \Phi(2)$$

$$= 1 - 0.9772 \quad \text{from tables}$$

$$= 0.0228$$

(b) $P(X < 95) = P(Z < -1) = P(Z > 1)$ by symmetry.

$$= 1 - P(Z \leqslant 1)$$

$$= 1 - \Phi(1)$$

$$= 1 - 0.8413 \quad \text{from tables}$$

$$= 0.1587$$

(c) $P(95 < X < 105)$ $= P(-1 < Z < 1)$

$$= 2P(0 < Z < 1)$$

$$= 2(\Phi(1) - \Phi(0))$$

$$= 2(\Phi(1) - 1/2)$$

$$= 2 \times 0.3413 \quad \text{from tables}$$

$$= 0.6826$$

(d) $P(105 < X < 115) = P(1 < Z < 3)$

$$= \Phi(3) - \Phi(1)$$

$$= 0.9987 - 0.8413 \quad \text{from tables}$$

$$= 0.1574$$

(e) $P(93 < X < 99)$ $= P\left(\dfrac{93 - 100}{5} < Z < \dfrac{99 - 100}{5}\right)$

$$= P(-1.4 < Z < -0.2)$$

$$= P(0.2 < Z < 1.4) \quad \text{by symmetry}$$

$$= \Phi(1.4) - \Phi(0.2)$$

$$= 0.9192 - 0.5793 \quad \text{from tables}$$

$$= 0.3399$$

Use of tables

If we use tables to obtain $\Phi(1.264)$ we can either approximate and find $\Phi(1.26)$ or for greater accuracy we can calculate as follows

$$\Phi(1.264) \simeq \Phi(1.26) + \frac{4}{10}(\Phi(1.27) - \Phi(1.26))$$

$$\simeq 0.8962 + \frac{4}{10}(0.0018)$$

$$\simeq 0.8969$$

Exercise 9.1

1. Use tables to find the value of
   (a) $\Phi(z)$
   (b) $1 - \Phi(z)$
   (c) $2(\Phi(z) - 0.5)$

   for (i) $z = 1.3$ (ii) $z = 2.1$ (iii) $z = 1.843$

2. If $\mu = 10$ and $\sigma = 3$ for the variate X, show that the standardised variate Z has mean 0 and standard deviation 1. Find also the value of Z corresponding to the value (a) $x = 15$, (b) $x = 6$, (c) $x = 5.4$.

3. If $\mu = 10$, $\sigma = 2$ for the Normal variate X find from tables

   (a) $P(X < 13)$ (b) $P(X < 15)$ (c) $P(X > 10)$
   (d) $P(X > 1)$ (e) $P(X < 9)$ (f) $P(X < 6)$
   (g) $P(7.5 < X < 10)$ (h) $P(10 < X < 12.5)$ (i) $P(7 < X < 11)$
   (j) $P(8.2 < X < 9.8)$

4. If X is $N(96, 3.4^2)$ find from tables

   (a) $P(X > 100)$ (b) $P(X < 100)$ (c) $P(X < 93 \text{ or } X > 99)$
   (d) $P(92 < X < 95)$ (e) $P(97 < X < 98)$

5. If Z is the standard Normal variate (that is Z is $N(0, 1)$) find the value z such that

   (a) $P(Z > z)$ $= 0.5,$ (d) $P(-1 < Z < z)$ $= 0.1,$
   (b) $P(Z < z)$ $= 0.75,$ (e) $P(-z < Z < z)$ $= 0.90.$
   (c) $P(Z < z)$ $= 0.975,$

6. If X is $N(5, 4)$ find the value x such that

   (a) $P(X < x)$ $= 0.5$,
   (b) $P(X < x)$ $= 0.75$,
   (c) $P(5 < X < x)$ $= 0.35$,
   (d) $P(-x < X < x)$ $= 0.80$,
   (e) $P(X < x)$ $= 0.05$.

7. IQ scores are $N(100, 225)$. Find the percentage of people expected to have a score
   (a) below 70, (b) between 80 and 120, (c) above 140.
   Find also the least score (to the nearest whole number) associated with the 20%
   of the population gaining the highest scores.

8. (a) Find the area of the Normal distribution included between the limits
   $\mu - 1.96\sigma$ and $\mu + 1.96\sigma$.
   (b) Just half of the area under the normal curve is included between the limits
   $\mu - \alpha\sigma$ and $\mu + \alpha\sigma$. Find α.

9. Tests on a certain brand of electric lamp indicate that the length of life is
   $N(1700, 400)$ measured in hours. Estimate the percentage of lamps which can be
   expected to burn (i) more than 1725 hours, (ii) less than 1690 hours, (iii) between
   1680 and 1710 hours.

10. Packages from a filling machine have a gross weight which is $N(1.01, 0.01)$, the
    units in kg. The label on the packet claims that the gross weight is 1 kg. Find the
    probability that a packet picked at random is underweight.

11. A Normal variate has a mean of 12.4 and variance 6.3. Find the probability that it
    will assume a value
    (a) larger than 15,
    (b) smaller than 16,
    (c) between 9.5 and 14.3.

12. An athlete finds that in the long jump his distances form a Normal distribution
    with mean 6.1 m and standard deviation 0.03 m.
    Calculate the probability that he will jump more than 6.17 m on a given
    occasion.
    Find the probability that three independent jumps will all be less than 6.17 m.
    What distance can he expect to exceed once in 500 jumps?

                                                                          AEB 1975

9.2 Further use of tables

Example 9.4
The distribution of heights in a large group of men of a similar age can be approximated
by a Normal distribution. Given that 10% have heights above 2 m and 18% of the group

have heights below 1.75 m, estimate the mean and standard deviation of heights of the whole population.

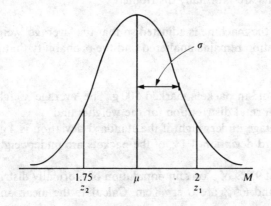

$$1 - \Phi(z_1) = 0.1 \Rightarrow \Phi(z_1) = 0.9$$

$$\Rightarrow \quad z_1 = 1.28 \quad \text{from tables.}$$

$$1 - \Phi(-z_2) = 0.18 \Rightarrow \Phi(-z_2) = 0.82$$

$$\Rightarrow -z_2 = 0.915$$

Also

$$\frac{2 - \mu}{\sigma} = z_1 = 1.28, \tag{1}$$

$$\frac{\mu - 1.75}{\sigma} = -z_2 = 0.915. \tag{2}$$

Adding,

$$\frac{0.25}{\sigma} = 2.195$$

$$\Rightarrow \sigma = 0.1139.$$

Substituting into (1),

$$\mu = 2 - 1.28 \times 0.1139$$

$$= 1.85.$$

So the mean of the population is 1.85 m with standard deviation 0.114 m (approx)

Exercise 9.2

1. Experience has shown that a certain machine produces rods which have lengths distributed Normally with standard deviation 1 cm. Find the mean length if 99% of the output has a length of 6 cm or less. Find also the proportion of the output with lengths between 5.8 cm and 5.9 cm.

2. Sacks of grain packed by an automatic machine have a mean weight of 56 kg. It is found that 10% of the bags are over 57 kg. Find the standard deviation assuming that the weights are Normally distributed.

3. In question 2, the machine is adjusted so that the 'average' weight is 55 kg. If the standard deviation remains unaltered find the probability that a bag is now over 57 kg.

4. A product is sold in packets marked 500 g. The 'average' weight is in fact 510 g. Assuming a Normal distribution for the weights find
   (a) the percentage underweight if the standard deviation is 4 g,
   (b) the standard deviation if 1% of the packets are underweight.

5. It is found that 9.8% of a certain population of Normally distributed lengths are under 40 cm and 6.5% are over 48 cm. Calculate the mean and variance of the distribution.

6. An automatic filling machine is known to operate with a standard deviation of 1.5 g. To what 'average filling' should the machine be set so that 95% of the packets are over 250 g.

7. A filling machine pours material into cartons labelled '50 g net'. If the total material delivered in 1000 fillings is 53.2 kg and the distribution is Normal with standard deviation 1.8 g what percentage is underweight?
   The manufacturers wish to reduce this percentage to 1%, to what 'average filling' should the machine be set?

8. The marks of candidates in an examination are $N(68, 100)$ and are standardised as scores which are $N(100, 225)$. Find
   (a) the score corresponding to a mark of 50,
   (b) the mark corresponding to a score of 110.

9. The numbers below represent the times that I took to travel to school on 10 consecutive days, given to the nearest minute

$$36 \quad 34 \quad 31 \quad 29 \quad 40 \quad 37 \quad 35 \quad 34 \quad 39 \quad 35$$

Calculate the mean and standard deviation of these times.
   If the time taken for my journey is Normally distributed with the mean and variance of the above data, what is the probability of my journey taking longer than 37 minutes?
   How much time should I allow myself for travelling to school if I will accept being late, at the most, one day in 10?

10. Explain what is meant by the quartiles of a frequency distribution. Find the quartiles of the Normal distribution with mean 0 and variance 1.
    In a test of 1000 students, 250 scored less than 40% and 250 scored more than 66%. Find the mean and variance of the scores.

11. Ball-bearings are produced with a diameter which is approximately Normally distributed with mean 6 mm and standard deviation 0.2 mm. A firm orders 1000 ball-bearings, with diameters within the limit 6 mm $\pm$ 0.1 mm. If the 1000 ball-bearings are selected at random, how many can be expected to satisfy the conditions?

12. A sample of 100 apples is taken from a load. The apples have the following distribution of sizes

Diameter to nearest cm	6	7	8	9	10
Frequency		11	21	38	17	13

Determine the mean and standard deviation of these diameters.

Assuming that the distribution is approximately normal with this mean and this standard deviation find the range of size of apples for packing if 5% are to be rejected as too small and 5% are to be rejected as too large.

o and c

13. Before joining the Egghead Society, every candidate is given an intelligence test which, applied to the general public, would give a normal distribution of IQs with mean 100 and standard deviation 20.

The candidate is not admitted unless his IQ, as given by the test, is at least 130.

Estimate the median IQ of the members of the Egghead Society, assuming that their IQ distribution is representative of that of the part of the population having IQs greater than, or equal to, 130.

What IQ would be expected to be exceeded by one member in ten of the society?

AEB 1976

9.3 The Normal approximation to the Binomial distribution

Suppose X is a discrete variate distributed as $B(n, p)$; suppose further that a relative frequency histogram is drawn to represent the probability distribution of X. The shape of such a histogram depends on the values of n and p.

If np is very small for large n, the distribution is skew. For example:

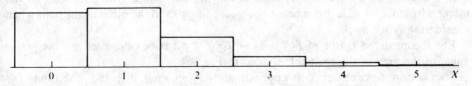

Fig. 9.5   A relative frequency histogram of $B(20, 0.05)$.

If p is close to 0.5 then the histogram representing the distribution is nearly symmetrical (actually symmetrical if $p = 0.5$), regardless of the value of n.

For example:

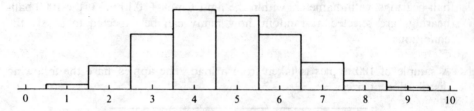

Fig. 9.6   A relative frequency histogram of $B(10, 0.45)$.

In fact, no matter what value p takes, as n is increased the distribution becomes more and more symmetrical. Further, it can be shown that whenever n is large and p is neither too large nor too small, not only is the distribution almost symmetrical, but in a sense yet to be explained, it is approximately Normally distributed. The approximation improves as n is increased though the approximation is satisfactory whenever

(i) $n > 30$,
(ii) $5 < np < n - 5$, i.e. $np > 5$, and $nq > 5$.

See Example 12.8 for a further discussion of this point.

Suppose n and p are such as to satisfy the above conditions then the relative frequency histogram representing $B(n, p)$ might be as in Fig. 9.7.

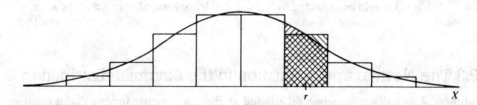

Fig. 9.7   A Normal approximation to a Binomial distribution.

If a Normal distribution is superimposed then it is found that the 'fit' is quite good. It is intuitively clear that the best Normal distribution to choose is one with the same mean and variance as $B(n, p)$, that is $N(np, npq)$. See p. 184.

In the diagram the horizontal scale is playing two roles. One consisting of the integers from 0 to n inclusive, when we consider $B(n, p)$ and the other as the reals when considering $N(np, npq)$.

For X considered as $B(n, p)$, $P(X = r) = {}^nCr\, q^{n-r}\, p^r$ can be calculated and is equal in value to the area of the rectangle shaded in Fig. 9.7.

If we choose to consider X as approximately $N(np, npq)$, then the probability (or area) to use is

$$P(r - \tfrac{1}{2} \leqslant x \leqslant r + \tfrac{1}{2}) = \Phi(r + \tfrac{1}{2}) - \Phi(r - \tfrac{1}{2})$$

Similarly if we wanted to find $P(X \geqslant r)$ then we would approximate this probability by calculating

$$P(X \geqslant r - \tfrac{1}{2}) \quad \text{in } N(np, npq).$$

Fig. 9.7 shows why the correction of 1/2 is necessary. It is due simply to the fact that we are moving from a discrete variate to a continuous variate. For this reason we often speak of the 'continuity correction'.

Example 9.5

A fair coin is tossed 400 times, find the probability of obtaining between 190 and 210 heads inclusive.

If we let X represent the number of heads obtained, then X is distributed as $B(400, 1/2)$.

The amount of work involved in finding $P(190 \leqslant X \leqslant 210)$ directly is daunting! Fortunately, since $n(= 400)$ is greater than 30 and $np\,(= 200)$ is neither too large nor too small, we can use the $N(200, 100)$ distribution as an approximation.

Now, if we think of X distributed as $N(200, 100)$, the corresponding probability to establish is

$$P(189.5 \leqslant X \leqslant 210.5) = P(-1.05 \leqslant Z \leqslant 1.05)$$
$$= 2(\Phi(1.05) - \tfrac{1}{2})$$
$$= 2 \times 0.8531 - 1$$
$$= 0.7062$$
$$\simeq 0.71$$

Note the saving in time in using an approximation!

Exercise 9.3

1. Find the probability of obtaining between 4 and 6 heads inclusive in ten tosses of a fair coin
   (a) using the Binomial distribution,
   (b) using the Normal approximation.

2. A student sits a multiple-choice examination, consisting of 40 questions each of which require the student to choose one of five possible answers. If the student simply guesses the answers to each question find the probability that he obtains
   (a) more than 10 correct answers,
   (b) more than 12 correct answers.

3. A fair die is tossed 120 times. Find the probability of obtaining fewer than 15 sixes.

4. A certain make of toy car is packed into cartons consisting of 500 cars. Past experience shows that about 1% of the cars are defective. Find the probability that there are more than three defective cars in a given carton.

5. If $\frac{1}{3}$ of the population vote Labour and 1000 investigators each interview 20 people, how many would report two or fewer Labour voters using
   (a) the Binomial distribution,
   (b) the Normal approximation.

6. A large number of students took an examination. The mean mark was 60 with a standard deviation of 10 marks, and 30 students had a mark of more than 70. Estimate the number of students who took the test.

7. The probability that a silicon chip will fail in less than 1000 hours of use is 0.3. Find the probability that of 100 such chips less than 60 are working after 1000 hours of continuous use.

8. If X is $B(40, 0.4)$ find the probability that
   (a) $X = 16$,
   (b) $12 < X < 20$.

9. Find the probability that in 120 tosses of a fair coin
   (a) between 40% and 60% will be heads,
   (b) 70% or more will be heads.

10. In Urbania selection for the Royal Flying Corps (RFC) is by means of an aptitude test based on a week's intensive military training. It is known that the scores of potential recruits on this test follow a Normal distribution with mean 45 and standard deviation 10.
    (a) What is the probability that a randomly chosen recruit will score between 40 and 60?
    (b) What percentage of the recruits is expected to score more than 30?
    (c) In a particular year 100 recruits take the test. Assuming that the pass mark is 50, calculate the probability that less than 35 recruits qualify for the RFC.

    AEB 1978

11. State necessary conditions for an observed quantity to follow the Binomial distribution. Give examples of two quantities you would expect to satisfy these conditions, giving your reasons and the parameters of the distributions.
    Jim is cracking nuts which have been chosen at random from a large collection of nuts of which 25% are bad. He cracks 100 nuts. (a) Write down an expression for the exact probability that he obtains 70 or more good nuts. (b) Use an approximation to derive a numerical value (correct to 2 DP) for this probability.

    OXFORD

12. A telephone exchange serves 2000 subscribers, and at any moment during the busiest period there is a probability of 1/30 for each subscriber that he will require a line. Assuming that the needs of subscribers are independent, write down an expression for the probability that exactly N lines will be occupied at any moment during the busiest period.
    Use the Normal distribution to estimate the minimum number of lines that

would ensure that the probability that a call cannot be made because all the lines are occupied is less than 0.01.

Investigate whether the total number of lines needed would be reduced if the subscribers were split into two groups of 1000, each with its own set of lines.

MEI

13. State the precise conditions under which the Normal distribution can be used as an approximation to the Binomial distribution.

A machine produces screws, some of which are known to be faulty. An inspection scheme is devised by taking random samples of 1000 screws from a large batch produced by the machine and noting the number x of faulty screws in the sample. If $x > 80$ the batch is rejected, but if $x \leqslant 80$ the batch is accepted. Find the probabilities of
 (i) accepting a batch containing 5% faulty screws,
 (ii) rejecting a batch containing 10% faulty screws.

MEI

Chapter 10 Expectation algebra revisited

10.1 Independent variates

It was stated in Chapter 5 that two events A and B are independent if the occurrence of one in no way affects the occurrence of the other. More formally, we state that the events A and B are independent if

$$P(A \text{ and } B) = P(A) \cdot P(B)$$

Suppose X and Y are two variates, then we shall say that the variates X and Y are independent if the values of X are in no way affected by the values obtained for Y. More formally, if x is any possible value of X and y is any possible value of Y then we say that the random variates X and Y are independent if

$$P(X = x \text{ and } Y = y) = P(X = x) \cdot P(Y = y)$$

The sum of two variates

If X and Y are two variates we define a new variate.

$$Z = X + Y$$

The values of Z are found by obtaining a value x (say) of X and a value y (say) of Y and adding these values together.

Similarly if we set $Z = 3X - Y$ then we obtain a value of Z by multiplying a value of X by 3 and subtracting a value of Y.

Suppose, for example, X and Y are *independent* variates with probability distributions as follows

x	1	2	3		y	0	1
p_r	$\frac{1}{4}$	$\frac{1}{2}$	$\frac{1}{4}$		p_r	$\frac{1}{2}$	$\frac{1}{2}$

We can form two tables. The first representing the possible sums and the second the probability that these particular sums occur.

+	0	1	y
1	1	2	
2	2	3	
3	3	4	
x			

(A)

p_r	0 $\frac{1}{2}$	1 $\frac{1}{2}$	y
1 $\frac{1}{4}$	$\frac{1}{8}$	$\frac{1}{8}$	
2 $\frac{1}{2}$	$\frac{1}{4}$	$\frac{1}{4}$	
3 $\frac{1}{4}$	$\frac{1}{8}$	$\frac{1}{8}$	
x			

(B)

From Table (B) notice that $P(X = 2 \text{ and } Y = 0) = \frac{1}{4}$.
This is because X and Y are independent and so

$$P(X = 2 \text{ and } Y = 0) = P(X = 2) \cdot P(Y = 0)$$

$$= \tfrac{1}{2} \times \tfrac{1}{2}$$

$$= \tfrac{1}{4}$$

and indeed this is how Table (B) was constructed.
Such a table is called a *joint probability distribution*. (We can construct joint probability distributions even if the variables are not independent.)

Now if $Z = X + Y$, note from Table (A) that the possible values of Z are 1, 2, 3, or 4 and using both tables we see, for example, that

$$P(Z = 3) = \tfrac{1}{4} + \tfrac{1}{8}$$

$$= \tfrac{3}{8}$$

Alternatively, without using the tables directly, we could argue

$$P(Z = 3) = P((X = 2 \text{ and } Y = 1) \text{ or } (X = 3 \text{ and } Y = 0))$$

$$= P(X = 2 \text{ and } Y = 1) + P(X = 3 \text{ and } Y = 0),$$

since the events are mutually exclusive

$$= P(X = 2) \cdot P(Y = 1) + P(X = 3) \cdot P(Y = 0),$$

since X and Y are independent.

$$= \frac{1}{2} \cdot \frac{1}{2} + \frac{1}{4} \cdot \frac{1}{2}$$

$$= \frac{3}{8}.$$

Combining these results we can obtain the probability distribution of Z as

z	1	2	3	4
p_r	$\frac{1}{8}$	$\frac{3}{8}$	$\frac{3}{8}$	$\frac{1}{8}$

10.2 The mean and variance of the sum of two independent variates

It is easy to calculate the expectation of the variates above as

$$E(X) = 2 \qquad E(Y) = \tfrac{1}{2} \qquad E(Z) = 2\tfrac{1}{2}$$

Now clearly, in this case, $E(Z) = E(X) + E(Y)$, where $Z = X + Y$. This result is now established more generally.

Statement 10.1

> Suppose X and Y are independent variates and $Z = X + Y$ then
> $$E(Z) = E(X) + E(Y)$$

For suppose X and Y have probability distributions as indicated below

X	x_1	x_2	x_3	...	x_n	Y	y_1	y_2	y_3	...	y_m
p_r	p_1	p_2	p_3	...	p_n	q_r	q_1	q_2	q_3	...	q_m

The probability distribution for Z can be written in the form of an array, for convenience.

Z	$x_1 + y_1$	$x_1 + y_2$	$x_1 + y_3$	...	$x_1 + y_m$
p_r	$p_1 q_1$	$p_1 q_2$	$p_1 q_3$		$p_1 q_m$
Z	$x_2 + y_1$	$x_2 + y_2$	$x_2 + y_3$	...	$x_2 + y_m$
p_r	$p_2 q_1$	$p_2 q_2$	$p_2 q_3$		$p_2 q_m$
Z	$x_n + y_1$	$x_n + y_2$	$x_n + y_3$	...	$x_n + y_m$
p_r	$p_n q_1$	$p_n q_2$	$p_n q_3$		$p_n q_m$

$E(Z)$ can be obtained by summing each of the products of the form $(x_i + y_j) p_i q_j$. Since

$$(x_i + y_j) p_i q_j = x_i p_i q_j + y_j p_i q_j$$

it is convenient to add the terms of the form $x_i p_i q_j$ by columns and those of the form $y_j p_i q_j$ by rows to obtain

$$+ q_1 (p_1 x_1 + p_2 x_2 + p_3 x_3 + \quad ... \quad + p_n x_n)$$
$$+ q_2 (p_1 x_1 + p_2 x_2 + p_3 x_3 + \quad ... \quad + p_n x_n)$$
$$+ \ \vdots$$
$$+ q_m (p_1 x_1 + p_2 x_2 + p_3 x_3 + \quad ... \quad + p_n x_n)$$
$$+$$
$$+ p_1 (q_1 y_1 + q_2 y_2 + q_3 y_3 + \quad ... \quad + q_m y_m)$$
$$+ p_2 (q_1 y_1 + q_2 y_2 + q_3 y_3 + \quad ... \quad + q_m y_m)$$
$$+ \ \vdots$$
$$+ p_n (q_1 y_1 + q_2 y_2 + q_3 y_3 + \quad ... \quad + q_m y_m)$$
$$= E(X)(q_1 + q_2 + q_3 + \quad ... \quad + q_m)$$
$$+ E(Y)(p_1 + p_2 + p_3 + \quad ... \quad + p_n)$$
$$= E(X) + E(Y), \quad \text{since} \quad q_1 + q_2 + \quad ... \quad + q_m = 1$$
$$\text{and} \quad p_1 + p_2 + \quad ... \quad + p_n = 1$$

It can be shown similarly that if X and Y are *independent* then
$$E(X \cdot Y) = E(X) \cdot E(Y)$$
It is also true that

> If X and Y are independent
>
> Var $(X + Y) = $ Var $(X) + $ Var (Y)

This last result is surprising but can be shown as follows

$$
\begin{aligned}
\text{Var } (X + Y) &= E[(X + Y)^2] - [E(X + Y)]^2 \\
&= E(X^2 + 2XY + Y^2) - [E(X) + E(Y)]^2 \\
&= E(X^2) + 2E(XY) + E(Y^2) \\
&\quad - E(X)^2 - 2E(X)E(Y) - E(Y)^2 \\
&= E(X^2) - E(X)^2 + 2(E(XY) - E(X)E(Y)) + E(Y^2) - E(Y)^2 \\
&= \text{Var } (X) + 0 + \text{Var } (Y) \\
&\qquad \text{since by independence } E(XY) = E(X) \cdot E(Y) \\
&= \text{Var } (X) + \text{Var } (Y)
\end{aligned}
$$

If X and Y are not independent then it remains true that
$$E(X + Y) = E(X) + E(Y)$$
but it is not necessarily true that
$$E(X \cdot Y) = E(X) \cdot E(Y)$$
or that
$$\text{Var } (X + Y) = \text{Var } (X) + \text{Var } (Y)$$

The above results apply also in the case where the variates X and Y are continuous.

Statement 10.2

> In summary:
>
> If X and Y are independent variates then
> $$E(X + Y) = E(X) + E(Y)$$
> $$E(X \cdot Y) = E(X) \cdot E(Y)$$
> $$\text{Var } (X + Y) = \text{Var } (X) + \text{Var } (Y)$$
> But if X and Y are not independent then we only know that
> $$E(X + Y) = E(X) + E(Y)$$

Example 10.1

Given that X and Y are independent variates express the following in terms of the mean and variance of X and Y

(a) $E(3X + 2Y)$ (c) Var $(2X + Y)$

(b) $E(X^2)$ (d) Var $(X - Y)$

The following results have already been established in Section 6.3.

$$E(aX) = aE(X)$$

$$\text{Var } (aX) = a^2 \text{ Var } (X), \text{ where } a \text{ is a constant}$$

(a) $E(3X + 2Y)$ $= E(3X) + E(2Y)$

$\qquad\qquad\qquad = 3E(X) + 2E(Y)$

(b) $E(X^2)$ $\qquad = \text{Var } (X) + E(X)^2,$ by definition

(c) Var $(2X + Y) = \text{Var } (2X) + \text{Var } (Y),$ by independence

$\qquad\qquad\qquad = 4 \text{ Var } (X) + \text{Var } (Y)$

(d) Var $(X - Y)$ $= \text{Var } (X + - Y)$

$\qquad\qquad\qquad = \text{Var } (X) + \text{Var } (- Y),$ by independence

$\qquad\qquad\qquad = \text{Var } (X) + \text{Var } (Y)$

This last result may at first sight be surprising, but Var $(- Y) = \text{Var } (-1 \cdot Y) = (-1)^2 \cdot \text{Var } (Y) = \text{Var } (Y)$. Though it is perhaps more obvious to note that changing the sign of each possible value of Y will not change the variance.

It is important to remember that, for *independent* variables X and Y:

Statement 10.3

$$\boxed{\text{Var } (X \pm Y) = \text{Var } (X) + \text{Var } (Y)}$$

Exercise 10.1

1. If the variate X represents the score obtained when an unbiased die is tossed and the variate Y represents the number of heads obtained when four unbiased coins are thrown find

   (a) $E(X)$, (b) $E(Y)$, (c) Var (X), (d) Var (Y), (e) $E(X + Y)$,
   (f) Var $(X + Y)$, (g) $E(2X - Y)$, (h) Var $(2X - Y)$.

   Write down without further calculation (i) $E(X^2)$, (ii) $E(Y^2)$.

2. If X represents the total score obtained when four fair dice are thrown and Y

represents the number of heads obtained when five unbiased coins are tossed, find
(a) Var (Y),   (b) $E(X + Y)$ (c) Var $(3X - 4Y)$

3. X and Y are independent variates each with mean μ and variances σ^2 and $2\sigma^2$ respectively. Write down in terms of μ and σ^2 the values of

(a) $E(X + 3)$,   (b) Var $(X + Y)$,   (c) Var $(X - Y)$,
(d) Var $(3X + 2Y)$,   (e) Var $(4X - Y - 2)$,   (f) Var $(aX + bY)$,

where a and b are constants.

4. Three dice are coloured white, red and blue respectively. After casting them a boy scores in the following way. To the 'white score' he adds twice the 'red score' and then subtracts the 'blue score'. Calculate the mean and variance of the final score.

5. A boy argues that

$$E(X + X) = E(X) + E(X)$$
$$= 2E(X)$$

and

$$\text{Var } (X + X) = \text{Var } (X) + \text{Var } (X)$$
$$= 2 \text{ Var } (X)$$

His friend agrees with the first result, but points out that

$$\text{Var } (X + X) = \text{Var } (2X)$$
$$= 4 \text{ Var } X.$$

Who made the error and why?

6. A Bernoulli trial can be modelled by a random variable X with probability distribution

x	0	1
p_r	q	p

where $X = 0$ indicates a failure and $X = 1$ indicates a success of the trial with probability p of success.
Calculate $E(X)$ and Var (X).
   If $X_1, X_2, \ldots, X_n$ are independent variates, each with the same probability distribution as X, write down the value of

$$E(X_1 + X_2 + \ldots + X_n)$$

and

$$\text{Var } (X_1 + X_2 + \ldots + X_n)$$

Explain why this gives us a proof of the facts noted in Chapter 7 concerning the mean and variance of $B(n, p)$.

7. The variate X has mean μ and variance σ^2 express $E((X + a)^2)$ in terms of μ and σ^2 where a is a constant.

10.3 Probability generating functions

So far we have met several probability distributions and have either calculated or simply stated the mean and variance. The two outstanding results yet to establish are the mean and variance of both the Binomial distribution $B(n, p)$ and the Geometric distribution. We have refrained from giving formal proofs earlier since now we can give an elegant derivation.

Suppose X is a discrete variate with probability distribution given by

x	x_1	x_2	x_3	$\ldots$	x_n
p_r	p_1	p_2	p_3	$\ldots$	p_n

Consider the expression $E(t^X)$ where t is a new variable (not a variate). We know that the value of this expression is written as

$$E(t^X) = p_1 t^{x_1} + p_2 t^{x_2} + p_3 t^{x_3} + \ldots + p_n t^{x_n}$$

N.B. $E(t^X)$ is a function of t and will thus give different values for different choices of t.
Now from this rather formal viewpoint we can establish some interesting results.

1. When $t = 1$ $E(t^X) = 1$

    This is surely obvious anyway but

    $$E(t^X) = p_1 t^{x_1} + p_2 t^{x_2} + \ldots + p_n t^{x_n}$$

    Set $t = 1$

    $$= p_1 + p_2 + \ldots + p_n$$

    $$= 1, \quad \text{by the fundamental definition of a probability distribution.}$$

2. $\dfrac{d}{dt}(E(t^X))\Big|_{t=1} = E(X)$

    The notation above means, differentiate the expression $E(t^X)$ with respect to t then substitute $t = 1$. We are to show that this is simply $E(X)$.

    $$E(t^x) = p_1 t^{x_1} + p_2 t^{x_2} + \ldots + p_n t^{x_n}$$

    $$\frac{d}{dt} E(t^X) = p_1 x_1 t^{x_1 - 1} + p_2 x_2 t^{x_2 - 1} + \ldots + p_n x_n t^{x_n - 1}$$

    setting $t = 1$

    $$\frac{d}{dt} E(t^X)\Big|_{t=1} = p_1 x_1 + p_2 x_2 + \ldots + p_n x_n$$

    $$= E(X) \quad \text{by definition.}$$

3. $\dfrac{d^2}{dt^2}(E(t^X))\Big|_{t=1} = E(X^2) - E(X)$

We have seen that

$$\frac{d}{dt}E(t^X) = p_1 x_1 t^{x_1-1} + p_2 x_2 t^{x_2-1} + \ldots + p_n x_n t^{x_n-1}$$

so, differentiating again,

$$\frac{d^2}{dt^2}(E(t^X)) = p_1 x_1 (x_1-1) t^{x_1-2} + p_2 x_2 (x_2-1) t^{x_2-2} + \ldots + p_n x_n (x_n-1) t^{x_n-2}$$

Setting $t = 1$

$$\begin{aligned}
\frac{d^2}{dt^2}(E(t^X))\Big|_{t=1} &= p_1 x_1 (x_1 - 1) + p_2 x_2 (x_2 - 1) + \ldots + p_n x_n(x_n - 1) \\
&= p_1 x_1^2 + p_2 x_2^2 + \ldots + p_n x_n^2 \\
&\quad - (p_1 x_1 + p_2 x_2 + \ldots + p_n x_n) \\
&= E(X^2) - E(X), \quad \text{by definition.}
\end{aligned}$$

We now have all the ingredients required for finding $E(X)$ and Var (X).

The function of t, $E(t^X)$, is called the probability generating function of X (pgf) and is often denoted by $G(t)$. Using this notation we collect together the results given above, and repeat the proofs

(1) $\qquad G(1) = 1$

since

$$\begin{aligned}
G(1) &= E(t^x)\big|_{t=1} \\
&= E(1) \\
&= 1
\end{aligned}$$

(2) $\qquad G'(1) = E(X)$ $\qquad\qquad G'(t) = \dfrac{d}{dt}(G(t))$

since

$$G'(t) = \frac{d}{dt}(E(t^X))$$

(the derivative of $G(t)$ with respect to t)

$$= E\left(\frac{d}{dt}t^X\right)$$

$$G'(1) = G'(t)\big|_{t=1}$$

$$= E(Xt^{X-1})$$

so

$$G'(1) = E(X)$$

(3) $G''(1) = E(X^2) - E(X)$

since

$$G''(t) = \frac{d^2}{dt^2}(E(t^X))$$

$$= E\left(\frac{d^2}{dt^2}(t^X)\right)$$

$$= E(X(X-1)t^{X-2})$$

$G''(t) = \frac{d^2}{dt^2}(G(t))$

this interchange is valid for all the situations we shall consider

so

$$G''(1) = E(X(X-1))$$

$$= E(X^2) - E(X)$$

We can now see that

Statement 10.4

$$\boxed{\begin{array}{c} E(X) = G'(1) \\ \text{Var } (X) = G''(1) - G'(1)^2 + G'(1) \end{array}}$$

The mean and variance of the Binomial distribution

Suppose X is distributed as $B(n, p)$ then

$$G(t) = E(t^X) = p_0 t^0 + p_1 t^1 + p_2 t^2 + \ldots + p_n t^n,$$

since X takes the integral values from 0 to n *inclusive*.

$$= {}^nC_0\, q^n\, p^0\, t^0 + {}^nC_1\, q^{n-1}\, p^1\, t^1 + \ldots + {}^nC_n\, q^0\, p^n\, t^n$$

$$= (q + pt)^n.$$

The reason this function $(q + pt)^n$ is called a probability generating function is that when expanded the probabilities with which X takes its values are given as the coefficients of the powers of t

Now

$$E(X) = G'(1)$$

$$= n(q + pt)^{n-1}_p \,|_{t=1}$$

$$= n(q + p)p$$

$$= np \quad \text{since } p + q = 1$$

Further

$$G''(1) = np^2\,(n-1)(q + pt)^{n-2}\,|_{t=1}$$

$$= np^2\,(n-1)$$

$$= n^2p^2 - np^2$$

and since

$$\text{Var } (X) = G''(1) - G'(1)^2 + G'(1)$$
$$= n^2p^2 - np^2 - n^2p^2 + np$$
$$= np - np^2$$
$$= np(1 - p)$$
$$= npq \quad \text{since } p + q = 1$$

The mean and variance of the Geometric distribution

$$G(t) = E(t^X)$$
$$= p_1 t^1 + p_2 t^2 + p_3 t^3 + \dots$$
$$= pt^1 + qpt^2 + q^2pt^3 + \dots,$$

using the usual notation. This is a geometric series with sum

$$\frac{pt}{1 - qt}, \quad \text{providing } |qt| < 1.$$

Thus

$$G'(t) = \frac{(1 - qt)p + qpt}{(1 - qt)^2} = \frac{p}{(1 - qt)^2}, \quad \text{so } G'(1) = \frac{1}{p}$$

$$G''(t) = \frac{2pq}{(1 - qt)^2}, \quad \text{so } G''(1) = \frac{2q}{p^2}$$

Now

$$E(X) = G'(1)$$
$$= \frac{1}{p}$$

$$\text{Var } (X) = G''(1) - G'(1)^2 + G'(1)$$
$$= \frac{2q}{p^2} - \frac{1}{p^2} + \frac{1}{p}$$
$$= \frac{2q - 1 + p}{p^2}$$
$$= \frac{q}{p^2}$$

10.4 The variance of the sum of two independent variates

This section may be omitted on the first reading.

Suppose $Z = X + Y$ where the variate X has mean μ_X and variance σ_X^2 and the variate Y has mean μ_Y and variance σ_Y^2

From the previous section we know that

$$E(Z) = E(X + Y)$$
$$= E(X) + E(Y)$$
$$= \mu_X + \mu_Y$$

Now consider

$$\text{Var }(Z) = \text{Var }(X + Y)$$
$$= E\left([(X + Y) - (\mu_X + \mu_Y)]^2\right), \quad \text{by definition}$$
$$= E\left([(X - \mu_X) + (Y - \mu_Y)]^2\right)$$
$$= E\left((X - \mu_X)^2 + 2(X - \mu_X)(Y - \mu_Y) + (Y - \mu_Y)^2\right)$$
$$= E\left((X - \mu_X)^2\right) + 2E\left((X - \mu_X)(Y - \mu_Y)\right) + E\left((Y - \mu_Y)^2\right)$$
$$= \sigma_X^2 + 2E\left((X - \mu_X)(Y - \mu_Y)\right) + \sigma_Y^2$$

Definition 10.1

> We define the *Covariance* of X and Y as
>
> $$\text{Cov }(X, Y) = E\left((X - \mu_X)(Y - \mu_Y)\right)$$

It is convenient to consider the *Product moment correlation coefficient*

$$\rho = \frac{\text{Cov }(X, Y)}{\sigma_X \sigma_Y}$$

as the value of ρ does not depend on the units X and Y are measured in. Methods of estimating ρ from random samples are discussed in Chapter 17.

In summary

$$\text{Var }(Z) = \text{Var }(X) + 2\text{ Cov }(X, Y) + \text{Var }(Y)$$
$$= \sigma_X^2 + 2\rho\,\sigma_X \sigma_Y + \sigma_Y^2.$$

An alternative form of Cov (X, Y) is

$$\text{Cov }(X, Y) = E\left((X - \mu_X)(Y - \mu_Y)\right)$$
$$= E(X \cdot Y - X\mu_Y - Y\mu_X + \mu_X \mu_Y)$$
$$= E(X \cdot Y) - \mu_Y E(X) - \mu_X E(Y) + \mu_X \mu_Y$$
$$= E(XY) - \mu_X \mu_Y$$
$$= E(X \cdot Y) - E(X) E(Y)$$

If the variates X and Y are *independent* we know from Section 10.2 that

$$E(X \cdot Y) = E(X) \cdot E(Y)$$

Thus

$$\text{Cov}\,(X,\,Y) = \rho = 0$$

It then follows in this case that

$$\text{Var}\,(Z) = \text{Var}\,X + \text{Var}\,Y,$$

agreeing with the result previously stated.

The converse is not true, that is if $\text{Cov}\,(X,\,Y) = 0$ it does not follow that X and Y are independent. See Exercise 10.2, question 3.

Exercise 10.2

1. The random variables x_1 and x_2 have means μ_1 and μ_2 and variances σ_1^2 and σ_2^2 respectively. The product-moment correlation coefficient between x_1 and x_2 is ρ. If $y = x_1 + x_2$ write down (i) the expected value of y, and (ii) the variance of y.

   A certain article is manufactured in two stages. The times, in minutes, to complete these two stages were observed for a large number of articles. It was found that the mean time for stage one was 24 min and the standard deviation of these times was 2 min. Similarly for stage two the mean time was 40 min and the standard deviation was 4 min. The correlation coefficient between the times for the two stages was 0.40. Using tables of the Normal distribution, estimate the proportion of items which will take more than 70 min to complete the two stages.

   MEI

2. X_1 and X_2 are independent variates each with mean μ and variance σ^2.

   Given that $Y_1 = aX_1 + bX_2$ and $Y_2 = bX_1 + aX_2$
   find

   (i) $\text{Var}\,(Y_1)$,
   (ii) $\text{Cov}\,(Y_1,\,Y_2)$.

3. The joint probability distribution of the variates X and Y is given below

		y			
		-1	0	1	
x	-1	$\frac{1}{16}$	$\frac{3}{8}$	$\frac{1}{16}$	$\frac{1}{2}$
	1	$\frac{3}{16}$	$\frac{1}{8}$	$\frac{3}{16}$	$\frac{1}{2}$
		$\frac{1}{4}$	$\frac{1}{2}$	$\frac{1}{4}$	

   Show that X and Y are not independent.
   Show further that $E(X) = E(Y) = E(XY) = \text{Cov}\,(X,\,Y) = 0$.

4. Show that $\text{Cov}\,(X,\,Y) = \text{Cov}\,(Y,\,X)$.

5. Show that $\text{Cov}(X, X + Y) = \text{Cov}(XX) + \text{Cov}(X, Y)$ and simplify $\text{Cov}(X + Y, X - Y)$.

6. Find the pgf of the variate X which represents the number of tosses of an unbiased die required before the first six occurs. Use this pgf to find (a) $E(X)$,   (b) Var (X).

7. Two players A and B toss two coins in turn. The first player to obtain two heads wins the game. If A starts find the probability that he wins the game.
   Assuming A wins the game, let X represent the number of attempts taken by A. Find the pgf of X and hence obtain its mean and standard deviation.

8. The probability that a discrete random variable X takes the value r is given by

$$P(X = r) = p_r, \quad r = 0, 1, 2, \ldots, n.$$

The probability generating function of X is

$$G(t) = \sum_{r=0}^{n} p_r t^r.$$

Derive the results

$$\mu = G'(1), \quad \sigma^2 = G''(1) + G'(1) - [G'(1)]^2,$$

where μ, σ^2 are the mean and the variance of X.
   If X is binomially distributed with

$$p_r = \binom{n}{r} p^r q^{n-r}, \quad r = 0, 1, 2, \ldots, n,$$

where $q = 1 - p$ and $0 < p < 1$, show that

$$G(t) = (q + pt)^n.$$

Use the above results for μ and σ^2 to deduce the mean and the variance of the above binomial distribution.
   Show that it is not possible for a binomial distribution to have a variance greater than its mean.

JMB

9. Find the probability generating function of a binomial distribution whose mean is 1.8 and whose variance is 0.99. If the random variable X has this binomial distribution
   (a) state the greatest possible value that X may take (with non-zero probability),
   (b) write down a numerical expression for the probability that a randomly observed value of X will be equal to 3. (Do not evaluate the expression.)

JMB

Chapter 11   The Poisson distribution

11.1 Some properties of the exponential function

Before introducing the Poisson distribution, some properties of the exponential function are revised.

Functions of the form

$$y = a^x$$
$$y = a^{-x},$$ where a is a positive constant

have graphs as indicated in Fig. 11.1.

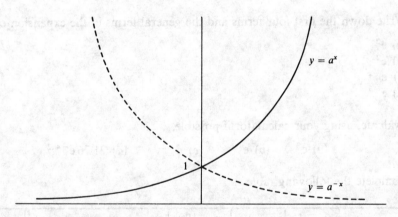

Fig. 11.1   A sketch of the graph of $y = a^x$ and $y = a^{-x}$ where the constant $a > 0$. Notice that $y = a^x$ is positive for all values of x.

Such functions have some interesting properties, for example,

$$\frac{dy}{dx} = Ka^x,$$

where K is a constant depending only on the value of a. If a is chosen to be $2.718\ldots$, then K turns out to be 1.

The number $2.718\ldots$ is called e and like π, $\sqrt{2}$ and many other numbers is irrational. Thus, for the exponential function

$$y = e^x,$$

we have

$$\frac{dy}{dx} = e^x.$$

It can be shown that

$$e^x = 1 + x + \frac{x^2}{2!} + \frac{x^3}{3!} + \frac{x^4}{4!} + \dots + \frac{x^r}{r!} + \dots,$$

where the summation on the right consists of an infinite number of terms.

This result can be used to evaluate e to as many decimal places as required since

$$e = e^1 = 1 + 1 + \frac{1^2}{2!} + \frac{1^3}{3!} + \frac{1^4}{4!} + \dots.$$

The sum of the first few terms gives quite a good approximation with the approximation improving as more and more terms are considered.

Exercise 11.1

1. Use the expansion of e^x given above to evaluate e correct to four decimal places.

2. Write down the first four terms and the general terms in the expansion of

   (a) e^2
   (b) e^{2x}
   (c) e^{-x}
   (d) e^{-a}.

3. Evaluate, using your calculator if possible,

         (a) e^3 (b) $e^{1.6}$ (c) $e^{-2.3}$ (d) $21.76\,e^{-4.2}$

4. Complete the following table

	$n = 1$	$n = 10$	$n = 100$	$n = $ your own choice	e^a
$a = 2$					
$a = -3$					
$a = $ your own choice					

by evaluating $\left(1 + \dfrac{a}{n}\right)^n$ for each of the values of a and n suggested. Also complete the last column.

The completed table will suggest that, no matter the value of a, as n increases the expression approaches the value e^a.

This result is written as

$$\lim_{n \to \infty}\left[\left(1 + \frac{a}{n}\right)^n\right] = e^a$$

11.2 The Poisson distribution

The most important distribution for discrete variates is the Binomial distribution. The second most important distribution for discrete variates is the Poisson distribution, introduced in this section. Though a useful distribution in its own right, we shall shortly see that it arises from the Binomial distribution in a natural way. Like the Binomial distribution it can also be approximated by the Normal distribution in certain circumstances.

Our first task is to define the probability distribution and calculate its mean and variance.

Definition 11.1

> A discrete random variable X defined for all non-negative integers follows the Poisson distribution $P(a)$ if
>
> $$P(X = r) = e^{-a}\frac{a^r}{r!},$$
>
> for each non-negative integer r, where a is a positive constant.

As with the Binomial distribution we can show that this is a probability distribution since

(i) $e^{-a}\dfrac{a^r}{r!} > 0$

for all suitable r since a is a positive constant and e^{-a} is always positive.

(ii) $\Sigma e^{-a}\dfrac{a^r}{r!} = e^{-a}\dfrac{a^0}{0!} + e^{-a}\dfrac{a^1}{1!} + e^{-a}\dfrac{a^2}{2!} + \ldots$

$$= e^{-a}\left(\frac{a^0}{0!} + \frac{a^1}{1!} + \frac{a^2}{2!} + \ldots\right)$$

$$= e^{-a}\left(1 + a + \frac{a^2}{2!} + \ldots\right)$$

$$= e^{-a} \cdot e^{a}$$

$$= 1,$$

as required for a probability distribution.
We now show that both the mean and variance of $P(a)$ is the parameter a.

Statement 11.1

> That is if X is $P(a)$ then
>
> $$E(X) = a$$
>
> $$\text{Var}\,(X) = a$$

The mean and variance of the Poisson distribution

The main task is to find the pgf of the Poisson variate X with parameter a, which can then be used to find $E(X)$ and Var (X).

In this case

$$G(t) = E(t^X)$$

$$= P_0 t^0 + p_1 t^1 + p_2 t^2 + \ldots$$

$$= e^{-a} \frac{a^0}{0!} t^0 + e^{-a} \frac{a^1}{1!} t^1 + e^{-a} \frac{a^2}{2!} t^2 + \ldots$$

$$= e^{-a} \left(\frac{a^0 t^0}{0!} + \frac{a^1 t^1}{1!} + \frac{a^2 t^2}{2!} + \ldots \right)$$

$$= e^{-a} e^{at}.$$

Thus

$$G'(t) = a e^{-a} e^{at}, \qquad \text{so} \quad G'(1) = a,$$

$$G''(t) = a^2 e^{-a} e^{at}, \qquad \text{so} \quad G''(1) = a^2,$$

so

$$E(X) = G'(1)$$

$$= a$$

$$\text{Var } (X) = G''(1) - G'(1)^2 + G'(1)$$

$$= a^2 - a^2 + a$$

$$= a$$

The Poisson distribution arises

(a) as an approximation to the Binomial distribution $B(n, p)$ which is satisfactory if

$$n > 50 \quad \text{and} \quad np \leqslant 5$$

The approximation improves when n is increased and p decreased in such a way that $np = a$ is a constant (cf. Section 11.3).

(b) When counting events which occur randomly in time or space.
  e.g. telephone calls, particle emissions and minute organisms well mixed in a fluid. The list is endless, but if the objects being counted tend to cluster, then the Poisson distribution is not applicable, e.g. frog spawn and traffic flow on a congested road.

Example 11.1

Apples are packed into boxes each containing 250. On average 0.6% of the apples are found to be bad. Find the probability that there will be more than two bad apples in a box.

  If X is a random variable representing the number of bad apples in a box and, assuming the apples are packed randomly and there has not been time for bad apples to affect others, then X is distributed as $B(250, 0.006)$.

Now since $np = a = 250 \times 0.006 = 1.5 \,(\leqslant 5)$ and $n = 250 \,(> 50)$ we can use $P(1.5)$ as a suitable approximation.

Thus

$$P(X = 0) = e^{-1.5}\frac{(1.5)^0}{0!} = 0.2231$$

$$P(X = 1) = e^{-1.5}\frac{(1.5)^1}{1!} = 0.3347$$

$$P(X = 2) = e^{-1.5}\frac{(1.5)^2}{2!} = 0.2510$$

So $\qquad\qquad P(X \leqslant 2) = \qquad\qquad\qquad 0.8088$

hence

$$P(X > 2) = 1 - P(X \leqslant 2)$$

$$= 0.1912$$

$$\simeq 0.19$$

So on average approximately one box in five will contain three or more bad apples. It is interesting to note that the use of the Binomial distribution gives $P(X > 2) = 0.1908$.

Example 11.2

A typist makes errors at random and on average makes two errors per page. Find the probability of making (a) no errors, (b) more than three errors on any given page.

On first reading we might suggest modelling this situation by using a Binomial distribution. A little thought shows this to be unlikely, for what could n be? In theory the typist could make any number of errors on a given page. However since the errors occur randomly we can use $P(2)$ as a model.

Now

$$P(X = 0) = e^{-2}\frac{2^0}{0!} = 0.1353$$

Also

$$P(X = 1) = e^{-2}\frac{2^1}{1!} = 0.2707$$

$$P(X = 2) = e^{-2}\frac{2^2}{2!} = 0.2707$$

$$P(X = 3) = e^{-2}\frac{2^3}{3!} = 0.1804$$

Thus
$$P(X \leqslant 3) = \overline{\quad\quad} \quad 0.8571$$

So
$$P(X > 3) = 1 - P(X \leqslant 3) = 0.1429$$

Hence
$$P(X = 0) \simeq 0.14, \quad \text{to 2DP.}$$
$$P(X > 3) \simeq 0.14, \quad \text{to 2DP.}$$

To calculate $P(X = r + 1)$ in terms of $P(X = r)$

It is worth noting that in general

$$P(X = r + 1) = \frac{e^{-a} a^{r+1}}{(r + 1)!}, \quad \text{when considering } P(a)$$

$$= \frac{a}{r + 1} \cdot \frac{e^{-a} a^{r}}{r!}$$

$$= \frac{a}{r + 1} P(X = r)$$

Thus in the example above, where we are considering $P(2)$ we have

$$P(X = 0) = 0.1353$$

and so

$$P(X = 1) = \frac{2}{1}(0.1353) = 0.2706$$

$$P(X = 2) = \frac{2}{2}(0.2706) = 0.2706$$

$$P(X = 3) = \frac{2}{3}(0.2706) = 0.1804,$$

which gives a considerable saving in time, particularly when using a calculator.

11.3 The Poisson distribution as a limiting form of the Binomial distribution

The Poisson distribution $P(a)$ arises from the Binomial distribution $B(n, p)$ as a limiting case as

$$n \to \infty$$
$$p \to 0$$

in such a way that $np = a$ remains constant. This can be shown as follows.

Consider the variate X with probability distribution $B(n, p)$ where

$$P(X = 0), \quad P(X = 1), \quad P(X = 2), \quad P(X = 3), \ldots, \quad P(X = r), \ldots$$

are given by the terms in the expansion of $(q + p)^n$:

$$q^n, \; nq^{n-1}p, \; \frac{n(n-1)}{2!}q^{n-2}p^2, \; \frac{n(n-1)(n-2)}{3!}q^{n-3}p^3, \ldots.$$

Replace p by $\dfrac{a}{n}$ and rewrite as

$$q^n, \; q^{n-1}a, \; \frac{1\left(1-\dfrac{1}{n}\right)q^{n-2}}{2!}a^2, \; \frac{1\left(1-\dfrac{1}{n}\right)\left(1-\dfrac{2}{n}\right)}{3!}q^{n-3}a^3, \ldots.$$

Now as n increases, terms of the form

$$1 - \frac{1}{n} \quad \text{and} \quad \left(1 - \frac{r}{n}\right)$$

tend to 1 whilst terms of the form

$$q^n = (1 - p)^n = \left(1 - \frac{a}{n}\right)^n$$

and

$$q^{n-r} = (1 - p)^{n-r} = \left(1 - \frac{a}{n}\right)^{n-r}$$

each tend to e^{-a} (see Exercise 11.1, question 4).

Thus for large enough n, we have, approximately,

$$e^{-a}, \quad e^{-a}a, \quad \frac{e^{-a}a^2}{2!}, \quad \frac{e^{-a}a^3}{3!}, \ldots, \quad \frac{e^{-a}a^r}{r!}, \ldots.$$

So X has approximately the Poisson distribution $P(a)$, and in the limit X will actually follow the Poisson distribution $P(a)$.

Using the fact that $B(n, p)$ has mean np and variance npq we give an alternative way to establish the mean and variance of the Poisson distribution (considered as an approximation to the Binomial distribution $B(n, p)$ when $n \to \infty$ as $p \to 0$, so that $np = a$ remains constant). Since

$$np = a$$

and

$$npq = a(1 - p)$$

$$= a\left(1 - \frac{a}{n}\right)$$

which tends to a as $n \to \infty$.

However, the reader should try not to think of the Poisson distribution simply as an approximation to the Binomial distribution, though this was how it was first derived by the French mathematician S. D. Poisson in 1837. It would still be an important distribution to study even without the connection. It arises whenever we are counting events which occur

(i) randomly in time or space,
(ii) independently,
(iii) uniformly, which means that the expected number of events in any interval is proportional to the length of the interval, and
(iv) such that the probability that two events occur simultaneously is zero.

These conditions are sufficient to derive the Poisson distribution mathematically.

Exercise 11.2

1. If the variate X has the probability distribution $P(a)$ write down

   (a) $P(X = 0)$,
   (b) $P(X = 5)$,
   (c) $P(X = r + 1)$.

2. If the variate X has the probability distribution $P(4)$ calculate

   (a) $P(X = 0)$,
   (b) $P(X < 4)$,
   (c) $P(X \geqslant 3)$.

3. It is known that 10% of the components produced in a particular factory are defective. Find the probability that in a sample of 10 components chosen at random exactly two will be defective using

   (a) the Binomial distribution,
   (b) the Poisson distribution.

4. If the probability that any one of 2000 patients will suffer from taking a certain drug is 0.002, find the probability that more than two patients would suffer if the drug were administered to every patient.

5. The average number of cars crossing a bridge is 300 per hour. Assuming that cars arrive at random, find the probability that more than three cars will cross in any given minute.

6. An office receives an average of 20 telephone calls per hour. Calculate the probability of there being

   (a) no calls,
   (b) less than three calls in any period of 5 minutes.

7. A small car hire firm owns five cars. On average the weekday demand is for two cars

and the demand is for three during the weekend. Find the probability of having to refuse a customer on

(a) Monday,
(b) Saturday,
(c) during the weekend,

assuming that cars are hired for a day at a time.

8. Calculate the mean and variance of the frequency distribution below. Assuming that the Poisson distribution is a suitable model calculate the expected frequencies using the calculated mean.

x	0	1	2	3	4	5
f	16	30	20	18	11	5

9. It was once suggested that a particular corp in the Prussian army was more prone to death by horse kicks than any other corp. Data was collected over a period of 20 years, noting the number of deaths due to horse kicks.

Deaths	0	1	2	3	4
Number of corps during a paritcular year	109	65	22	3	1

Use the mean of the frequency table to calculate the theoretical frequency distribution using the Poisson distribution as a model. What deductions can be made?

10. A caretaker notes that on average he replaces three light bulbs each month in the school in which he works. Find the probability that during a given month he needs to replace
(a) no bulbs,
(b) more than four bulbs.
Calculate the number of bulbs he should keep in stock in order to ensure that the probability that he runs out of bulbs during any given month is less than 0.01.

11. A typist makes an average of one error every three pages of typescript. Find the probability that she makes her first error on the tenth page.

12. The number of road accidents at a certain traffic roundabout has been found to follow a Poisson distribution with a mean of 0.8 accidents per week. Calculate, correct to 2 significant figures, the probabilities that (i) there will be at least 2 accidents in a particular week, (ii) there will be exactly 3 accidents in a particular three-week period.

    After road works have been carried out, the traffic is observed for a period of 8 weeks and in that time only 1 accident occurs. Calculate the probability of at most 1 accident in 8 weeks if the mean is still 0.8 accidents per week. Comment briefly on this result.

CAMB.

13. (a) If X is $P(a)$ and $Y = X + 1$ find the pdf of Y and write down the mean and variance of Y.

(b) If X is $P(a)$ and Z takes values 1, 2, 3, such that $P(Z = r) = kP(X = r)$, for some constant k, obtain the pdf of Z and obtain the mean and variance of Z.

11.4 The Normal approximation of the Poisson distribution

It can be shown that if the mean of the Poisson distribution $P(a)$ is large enough, say with $a > 25$, then the Normal distribution $N(a, a)$ can be used as an approximation to $P(a)$. In this case, of course, we must remember the continuity correction of 1/2, due to moving from the discrete variable to a continuous variable. This is for just the same reason as we outlined when considering the Normal distribution as an approximation to the Binomial distribution in Section 9.3.

The diagram in Fig. 11.2 indicates the situation with regard to the various approximations available and the conditions under which they apply.

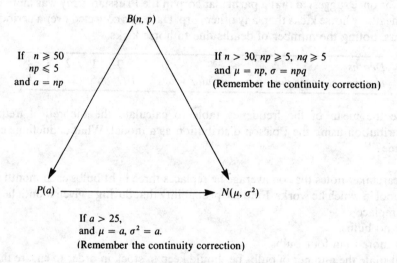

Fig. 11.2   The approximations to some distributions.

Example 11.3

Certain books are printed with an average of 50 errors per book. Find the probability that a book chosen at random will have more than 60 errors.

If we let X represent the number of errors in a book then it is reasonable that X should follow the Poisson distribution with mean 50. It is fortunate that the mean is greater than 25 since we can save a great deal of time by using the Normal approximation $N(50, 50)$.

In view of the continuity correction we wish to find

$P(X > 60.5)$

$= P\left(Z > \dfrac{60.5 - 50}{7.07}\right)$

$= P(Z > 1.485)$

$= 1 - P(Z < 1.485)$

$= 1 - \Phi(1.485)$

$= 1 - 0.9312$

$= 0.0688$

$\simeq 0.07$

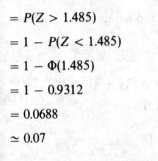

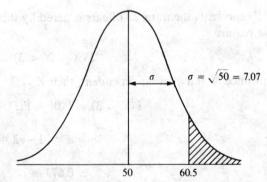

$\sigma = \sqrt{50} = 7.07$

50   60.5

11.5 The sum of two Poisson variates

We will show that if X is $P(a)$ and Y is $P(b)$, where the variates X and Y are independent then $Z = X + Y$ is $P(a + b)$.

The generating function for a variate which follows the Poisson distribution with mean a is $e^{-a} e^{at}$, where t is a dummy variable.

The generating function for Z is

$$E(t^Z) = E(t^{X+Y})$$

$$= E(t^X \times t^Y)$$

$$= E(t^X) \times E(t^Y),$$
since X and Y are independent

$$= e^{-a} e^{at} e^{-b} e^{bt}$$

$$= e^{-(a+b)} e^{(a+b)t},$$

which shows that Z is a Poisson variate with mean $(a + b)$.

Example 11.4

A car distributor supplies two particular garages which order cars independently each week with frequency following Poisson distribution with means 5/3 and 1/3 cars respectively.

Find the probability that in any given week less than three cars are ordered by
(a) the first garage,
(b) the two garages together.

(a) If we let X be a variable representing the number of cars ordered by the first garage, then X is $P(\frac{5}{3})$ approximately.

We require

$$P(X < 3) = P(0) + P(1) + P(2)$$

$$= e^{-5/3}\left(1 + \tfrac{5}{3} + \frac{(\frac{5}{3})^2}{2}\right)$$

$$\simeq 0.766.$$

(b) If Y represents the number of cars ordered by the second garage, then Y is $P(\frac{1}{3})$.
We require

$$P(X + Y < 3)$$

and since X and Y are independent, then $Z = X + Y$ is $P(\frac{5}{3} + \frac{1}{3})$ i.e. $P(2)$. Now

$$P(Z < 3) = P(0) + P(1) + P(2)$$

$$= e^{-2}\left(1 + 2 + \frac{2^2}{2}\right)$$

$$\simeq 0.677.$$

Exercise 11.3

1. The mean number of bacteria per millilitre of liquid is known to be 3. Find the probability that
   (a) a sample of 1 ml contains more than three bacteria,
   (b) a sample of 10 ml contains less than 25 bacteria,
   (c) two samples, each of 1 ml, contain no bacteria,
   (d) exactly two of three samples of 1 ml each, contain at least one bacterium.

2. In a certain factory the number of injuries per working week was recorded during a 48 week year.

Number of injuries	0	1	2	3
Number of weeks	28	16	3	1

   Calculate the mean and variance of the above frequency distribution. Fit a theoretical Poisson distribution using the same mean as the data. What evidence do you have that the Poisson model is suitable, and what does this suggest about the manner in which the accidents occur?
   Find the probability that during any given week there will be more than three such accidents.

3. If X is a random variable distributed as $P(2)$ find the least value V such that $P(X \geqslant V) < 1/2$.

4. Under which conditions do the following apply
   (a) the Binomial distribution,
   (b) the Poisson distribution,
   (c) the Poisson approximation to the Binomial distribution,
   (d) the Normal approximation to the Poisson distribution.

5. Five per cent of the clothes produced in a certain factory are substandard. Sixty items are chosen at random and sent for inspection to a prospective purchaser. Find the probability that less than three items are found to be substandard.

6. I receive, on average, four letters each weekday. Find the probability that I receive no letters
   (a) on Monday,
   (b) on just one of the five weekdays in a particular week,
   (c) during the whole of one week.

7. The number of monthly breakdowns of my car follows a Poisson distribution with mean 0.2. Find the probability that in a six-month period the car will run trouble free.

8. A hire company has two electric lawnmowers which it hires out by the day. The number of demands per day for a lawnmower has the form of a Poisson distribution with mean 1.50. In a period of 100 working days, how many times do you expect

   (i) neither of the lawnmowers to be in use,
   (ii) some requests for the lawnmowers to have to be refused?

   If each lawnmower is to be used an equal amount, on how many days in a period of 100 working days would you expect a particularly lawnmower not to be in use?

   MEI

9. A shopkeeper has two shops supplied from a central store. Each shop orders complete boxes of a particular product when required. The number of boxes requested per week for the two shops are independent and have a Poisson distribution with means 4/5 and 1/5 respectively. Find the probability that two or more boxes are requested from the store in a week (i) by the first shop, (ii) by the second shop, (iii) altogether.

10. Define the Poisson distribution. State its mean and variance. State under what circumstances the normal distribution can be used as an approximation to the Poisson distribution.

    Readings, on a counter, of the number of particles emitted from a radioactive source in a time T seconds have a Poisson distribution with mean $250\,T$. A ten-second count is made. Find the probabilities of readings of (i) more than 2600, (ii) 2400 or more.

    JMB

11. Define a Poisson distribution giving the general probability formula; state the relationship between its mean and variance.

    The distribution of the number of vehicles observed passing under a motorway bridge in 100 successive intervals of 12 seconds, at a time when the traffic flow is 720 vehicles per hour, may be considered to be Poissonian.

    During how many of the intervals would one expect
    (i) no cars to have passed,
    (ii) 2 cars to have passed,
    (iii) more than 2 cars to have passed?

    Find the smallest value of N such that the probability of at most N cars in an interval is greater than 0.999.

    AEB 1978

12. The number of telephone calls received at a switchboard in any time interval of length T minutes has a Poisson distribution with mean $\frac{1}{2}T$. The operator leaves the switchboard unattended for five minutes. Calculate to three decimal places the probabilities that there are
   (i) no calls,
   (ii) four or more calls in her absence.
   Find to three significant figures the maximum length of time in seconds for which the operator could be absent with a 95% probability of not missing a call.

JMB

13. In a large town, one person in 80, on the average, has blood of Type X. If 200 blood donors are taken at random, find an approximation to the probability that they include at least five persons having blood of type X.
   How many donors must be taken at random in order that the probability of including at least one donor of Type X shall be 0.9 or more?

AEB 1976

14. The number of oil tankers arriving at a port between successive high tides has a Poisson distribution with mean 2. The depth of the water is such that loaded vessels can enter the dock area only on the high tide. The port has dock space for only three tankers, which are discharged and leave the dock area before the next tide. Only the first three loaded tankers waiting at any high tide go into the dock area; any others must await another high tide.
   Starting from an evening high tide after which no ships remain waiting their turn, find (to three decimal places) the probabilities that after the next morning's high tide
   (i) the three dock berths remain empty,
   (ii) the three berths are all filled.
   Find (to two decimal places) the probability that no tankers are left waiting outside the dock area after the following evening's high tide.

JMB

Revision exercise B

1. A certain mass-produced component is acceptable if its length lies between 6.34 cm and 6.66 cm. It was found that approximately 9% were too short and 4% were too long. Assuming that the lengths are Normally distributed, estimate the mean and variance of the distributions and hence estimate the proportion which have lengths less than 6.4 cm.

2. It was established over a long period that there were on average 11 major accidents per year on a certain stretch of motorway. Assuming these accidents occur at random during the course of a year, estimate the probability that
   (a) there were no accidents in a given month,
   (b) there were no accidents in a given two month period.

3. Each week a random sample of 100 components is inspected to find the number of defective components. The results of 100 such samples are recorded below

Number of defectives	0	1	2	3	4	5	6 or more
Number of samples	19	30	30	17	3	1	0

Use this table to estimate the mean and variance of the number of defectives per sample.

Use an appropriate model to estimate the probability that a further sample of 100 components is found to have six or more defectives.

4. If the independent variates X and Y are chosen at random from the digits $0, 1, \ldots, 9$. find the expected value and the variance of (a) $X + Y$, (b) $3X - 2Y$.

5. Define the Binomial distribution, explaining clearly the symbols used.

A gun engages a target at a range at which the chance of a direct hit with any one round is $1/5$. How many rounds must be fired to give a 90% probability of at least one hit?

If, in fact, five rounds are fired and, failing a direct hit in these five rounds, a further five rounds are fired, find the chance of two or more direct hits in all.

JMB

6. A number of different types of fungi are distributed at random in a field. Eighty per cent of these fungi are mushrooms, and the remainder are toadstools. Five per cent of the toadstools are poisonous. A man, who cannot distinguish between mushrooms and toadstools, wanders across the field and picks a total of 100 fungi. Determine, correct to 2 significant figures, using appropriate approximations, the probability that the man has picked
 (i) at least 20 toadstools,
 (ii) exactly two poisonous toadstools.

CAMB

7. For each pupil arriving late at a school, the number of minutes late (to the nearest minute) is recorded. For 200 latecomers the times are as follows:

Minutes late	0–2	3–7	8–12	13–17	18–27
Frequency	80	75	25	10	10

Draw a histogram to represent the data. State (with a brief explanation) whether or not the histogram suggests that the distribution of lateness times is approximately normal.

Calculate estimates of the mean and standard deviation of the number of minutes late for these pupils.

JMB

8. The following table gives the cumulative frequency distribution of the masses x in kilograms of a group of 200 eighteen-year-old boys.

x	30	35	40	45	50	55	60	65	70	75	80	85	90	95
Number with mass less than x	0	1	4	11·	25	47	79	114	146	171	187	195	198	200

Draw a cumulative frequency graph and from this estimate the median.

Compile a frequency distribution from the data and hence estimate the mean and standard deviation of the sample, State a well known probability distribution which you would expect to fit such data.

JMB

9. An experiment consists of tossing a coin twice. Let A be the event that the first toss produces a head, B be the event that the second toss produces a head and C be the event that either two heads or two tails are obtained.

Find the following probabilities

$$P(A), \ P(B), \ P(A \cap C), \ P(A|C), \ P(C|B)$$

10. X is normally distributed with mean and standard deviation both equal to k units. Find $(X > 0)$, and if $k = 0.5$ evaluate $P(1 < X < 2)$.

It is desired to find the value of k which will maximise $P(1 < X < 2)$. Evaluate this probability for various values of k between 0.7 and 1.0, and hence find k correct to one decimal place.

CAMB

11. A count was made of the number of red blood corpuscles in each of the 64 compartments of a haemocytometer with the following results:

Number of corpuscles	2	3	4	5	6	7	8
Frequency	1	5	4	9	10	10	8
Number of corpuscles	9	10	11	12	13	14	
Frequency	6	4	3	2	1	1	

Estimate the mean and variance of the number of red blood corpuscles per compartment. Explain how the values you have obtained support the view that these data are a sample from a Poisson population.

Write down an expression for the theoretical frequency with which compartments containing 5 red blood corpuscles should be found, assuming this to be obtained from a Poisson population with mean 7. Evaluate this frequency to two decimal places, taking $e^{-7} = 9.119 \times 10^{-4}$.

MEI

12. The table gives the *cumulative* frequency distribution of the weights of a group of 800 eighteen-year-old women applicants to a large organisation.

Weight in kg not greater than	20	25	30	35	40	45	50
Cumulative frequency	0	1	13	44	107	198	336
Weight in kg not greater than	55	60	65	70	75	80	85
Cumulative frequency	481	611	692	745	779	791	800

(i) Plot the data on a cumulative frequency graph and estimate the median weight.

(ii) Derive a frequency distribution and hence estimate the mean and variance of the weights.

(iii) Assuming this group to be an appropriate random sample, use the Normal distribution to estimate the percentage of eighteen-year-old women with weights over average weight but less than 60 kg.

MEI

13. (a) A, B and C represent 3 events. If $A \cap B$ is the event that both A and B occur and $p(B|A)$ is the probability that B occurs given that A has already occurred, show that

$$p(A \cap B) = p(A)\, p(B|A).$$

Deduce, or show otherwise, that

$$p(A \cap B \cap C) = p(A)\, p(B|A)\, p(C|A \cap B).$$

(b) An athlete aims to measure his fitness by subjecting himself to a sequence of 3 physical tests, the completion of each test in a specified time being classed by him as a 'pass'. The probability that he passes the first test in the sequence is p, but the probability of passing any subsequent test is half the probability of passing the immediately preceding test. Show that, if the probability of passing all 3 tests is 1/216, the value of p is 1/3. Hence find the probabilities

(i) that he fails all the tests;

(ii) that he passes exactly 2 of the 3 tests.

MEI

14. In producing 500 tins of polish, a manufacturer has to satisfy two conditions as to the quantity contained in a tin. These are firstly that at least 97.5% of the output have masses exceeding 500 g and secondly that at least 99.9% of the output have masses exceeding 495 g. The masses satisfy a normal distribution.

(a) If the mean mass is 501 g, calculate the greatest possible standard deviation such that both conditions are satisfied.

(b) If the standard deviation of the masses is 0.5 g, calculate the least value of the mean such that both conditions are satisfied.

There is a loss of weight in time due to evaporation. After a two-year shelf life the loss is 5% with negligible variation. So that the goods sold in the shops shall meet the conditions as to quantities, the manufacturer makes the mean mass at the time of production 540 g. Calculate to two decimal places the greatest value of the standard deviation at manufacture such that both conditions still hold when the polish is sold after two years.

JMB

15. A trial may have two outcomes, success or failure. If in n such independent trials, the probability p of a success remains constant from trial to trial, write down the probability of r successes in the n trials.

When two friends A and B play chess, the probability that A wins any game is 2/5, and if A does not win the game, the probabilities then of B winning and of a draw are equal. In the course of an evening they play four games. Calculate the probabilities (i) that A does not win a game, (ii) that he wins more than two games.

If it is known that A has won exactly two of these four games, write down the probability distribution of the number of games that B has won.

Calculate the probability that A wins more games than B when four games are played.

JMB

16. Derive the mean and variance of the Binomial distribution.

It is found at a restaurant that, on the average, one in eight of the parties who book tables fail to appear. The restaurant has 48 tables and on one evening had accepted bookings for 52 tables. What is the probability that all the parties who appear can be accommodated?

AEB 1976

17. Fergus Lightfingers, an educated thief, has broken into a house and has come upon a large safe. Lying upon the table is a bunch of k similar keys, only one of which will open the safe. He considers two possible strategies:
   (i) randomly select keys, one at a time *without replacement,* until successful;
   (ii) randomly select keys, one at a time *with replacement,* until successful.

Let N, a random variable, be the number of keys including the successful one tried by Fergus in order to open the safe. Derive the probability distribution of N for both strategies. Calculate the expected value of N in each case.

Use your statistical judgement in order to select the strategy that Fergus should use.

(You may assume that $\sum\limits_{n=1}^{\infty} nx^{n-1} = 1/(1-x)^2$ for $|x| < 1$)

AEB 1978

18. Gnat larvae are distributed at random in pond water so that the number of larvae contained in a random sample of $10\,cm^3$ of pond water may be regarded as a random variable having a Poisson distribution with mean 0.2. Ten independent random samples, each of $10\,cm^3$, of pond water, are taken by a zoologist.

Determine (correct to three significant figures)
(a) the probability that none of the samples contain larvae,
(b) the probability that one sample contains a single larva and the remainder contain no larvae,
(c) the probability that one sample contains two or more larvae and the remainder contain no larvae,
(d) the expectation of the total number of larvae contained in the ten samples,
(e) the expectation of the number of samples containing no larvae.

CAMB

Chapter 12   Elementary sampling theory

12.1 Introduction

Let X be a continuous random variable with mean μ and variance σ^2. Suppose the value x is chosen at random. By which is meant an experiment is performed or a game played or a trial conducted or an object is measured, and thus the value of x is established. To say that x is chosen at random is to imply that there is no bias. In other words, if the values x_1, x_2, x_3, ..., x_n are established (chosen at random) then the histogram representing this data will approximate to the pdf of X with the approximation improving as more and more items are chosen.

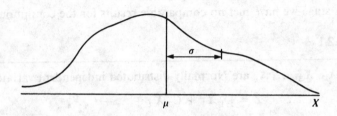

Establishing such values of X is called *random sampling* (from the population X). The population of values *obtained by sampling* can be thought of as a variate in its own right, say X_1. X_1 will be distributed in exactly the same way as X and so, in particular, will have the same mean and variance as X.

Suppose now that two values X_1 and X_2 of X are established independently and at random. Each of the values X_1 and X_2 can be viewed as random variables distributed in exactly the same way as X. What can be said of the variate Z where

$$Z = X_1 + X_2?$$

There are two results which are easy to obtain

$$E(Z) = E(X_1 + X_2)$$
$$= E(X_1) + E(X_2)$$
$$= \mu + \mu$$
$$= 2\mu$$
$$\text{Var}\,(Z) = \text{Var}\,(X_1 + X_2)$$
$$= \text{Var}\,(X_1) + \text{Var}\,X_2, \quad \text{by independence}$$
$$= \sigma^2 + \sigma^2$$
$$= 2\sigma^2.$$

It is generally quite difficult to obtain the pdf of Z, though in some cases the problem can be resolved.

Example 12.1
Suppose the discrete variate X takes values 0, 1 with equal probability. If $Z = X_1 + X_2$, where X_1 and X_2 are independent random values of X, then the probability distribution of Z is given by

z	0	1	2
p_r	$\frac{1}{4}$	$\frac{1}{2}$	$\frac{1}{4}$

the probabilities are calculated using the independence of X_1 and X_2.

Example 12.2
We have already seen that if X and Y are $P(a)$ and $P(b)$ respectively then $Z = X + Y$ is $P(a + b)$ provided X and Y are independent.

Until this stage we have met no comparable results for the continuous variates.

Statement 12.1

> If $X_1, X_2, X_3, \ldots, X_n$ are Normally distributed independent variates then
>
> $$Z = C_1 X_1 + C_2 X_2 + \ldots + C_n X_n$$
>
> is also Normally distributed where the $C_1, C_2, C_3, \ldots, C_n$ are constants.
> This is an extremely important result though it is difficult to prove at this stage.

Example 12.3
If X is $N(3, 9)$ and Y is $N(5, 16)$ and $Z = X + Y$, where the variates X and Y are independent, then Z is $N(8, 25)$.

Example 12.4
X is $N(\mu, \sigma^2)$ and X_1, X_2 are each independent random observations of X. Describe the distributions of

(a) $Z_1 = X_1 + X_2$ (c) $Z_3 = 2X_1$

(b) $Z_2 = X_1 - X_2$ (d) $Z_4 = 5X_1 + 2X_2$

(a) Since $Z_1 = X_1 + X_2$

$$E(Z_1) = E(X_1 + X_2) \qquad\qquad \text{Var}(Z_1) = \text{Var}(X_1 + X_2)$$
$$= E(X_1) + E(X_2) \qquad\qquad = \text{Var}(X_1) + \text{Var}(X_2)$$
$$= \mu + \mu \qquad\qquad\qquad\qquad = \sigma^2 + \sigma^2$$
$$= 2\mu \qquad\qquad\qquad\qquad\quad = 2\sigma^2$$

So Z_1 is $N(2\mu, 2\sigma^2)$

(b) Since $Z_2 = X_1 - X_2$

$$E(Z_2) = E(X_1 - X_2) \qquad\qquad \text{Var}\,(Z_2) = \text{Var}\,(X_1 - X_2)$$
$$= E(X_1) - E(X_2) \qquad\qquad = \text{Var}\,(X_1) + \text{Var}\,(X_2)$$
$$= \mu - \mu \qquad\qquad\qquad\qquad = \sigma^2 + \sigma^2$$
$$= 0 \qquad\qquad\qquad\qquad\qquad = 2\sigma^2$$

So Z is $N(0, 2\sigma^2)$

(c) Since $Z_3 = 2X_1$

$$E(Z_3) = E(2X_1) \qquad\qquad \text{Var}\,(Z_3) = \text{Var}\,(2X_1)$$
$$= 2E(X_1) \qquad\qquad\qquad = 4\,\text{Var}\,(X_1)$$
$$= 2\mu \qquad\qquad\qquad\qquad = 4\sigma^2$$

So Z is $N(2\mu, 4\sigma^2)$.

This result sometimes confuses students because they note that $Z_1 = X_1 + X_2$ has variance $2\sigma^2$ and $Z_3 = 2X_1 = X_1 + X_1$ has variance of $4\sigma^2$. The reason is that though X_1 or X_2 may on rare occasions be a long way from the mean, it will be less common for both to be a long way from the mean at the same time. So Z_1 will be a long way from the mean rather less often than Z_3.

(d) It is similarly easy to show that $Z_4 = 5X_1 + 2X_2$ is $N(7\mu, 29\,\sigma^2)$.

12.2 The distribution of the sample means obtained from a Normal population

Suppose X is $N(\mu, \sigma^2)$ and a random sample of size n is taken:

$$X_1, X_2, X_3, \ldots, X_n$$

then

$$\bar{X} = \frac{X_1 + X_2 + X_3 + \ldots + X_n}{n}$$

$$= \frac{1}{n}X_1 + \frac{1}{n}X_2 + \ldots + \frac{1}{n}X_n$$

the mean of the random sample can be viewed as a variate in its own right.

$$E(\bar{X}) = E\left(\frac{1}{n}X_1 + \frac{1}{n}X_2 + \ldots + \frac{1}{n}X_n\right)$$

$$= \frac{1}{n}\mu + \frac{1}{n}\mu + \ldots + \frac{1}{n}\mu$$

$$= \mu$$

$$\text{Var } \bar{X} = \text{Var}\left(\frac{1}{n}X_1 + \frac{1}{n}X_2 + \ldots + \frac{1}{n}X_n\right)$$

$$= \frac{1}{n^2}\sigma^2 + \frac{1}{n^2}\sigma^2 + \ldots + \frac{1}{n^2}\sigma^2 \quad \text{by independence}$$

$$= \frac{\sigma^2}{n}.$$

So $\bar{X}$ is $N\left(\mu, \frac{\sigma^2}{n}\right)$

Diagrammatically

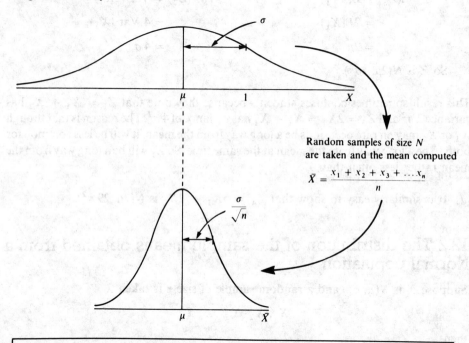

Random samples of size N
are taken and the mean computed

$$\bar{X} = \frac{x_1 + x_2 + x_3 + \ldots x_n}{n}$$

> The distribution of the sample means $\bar{X}$ is a most important distribution. Its standard deviation is often called the standard error of the mean.

Example 12.5
If X is $N(10, 9)$ and a random sample of size 16 is taken and found to have a mean $\bar{X}$, find $P(9 \leqslant \bar{X} \leqslant 11)$.

$$P(9 \leqslant \bar{X} \leqslant 11) = P(-1.33 \leqslant Z \leqslant 1.33)$$

$$= 2(\Phi(1.33) - \tfrac{1}{2})$$

$$= 0.818.$$

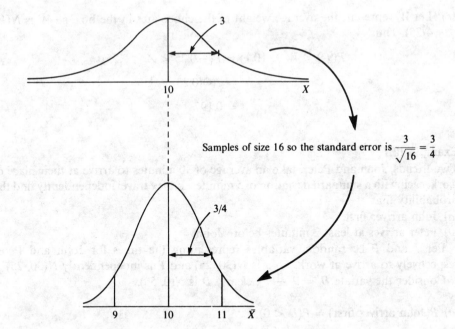

Samples of size 16 so the standard error is $\dfrac{3}{\sqrt{16}} = \dfrac{3}{4}$

Example 12.6

Chocolates are produced with weight W g where W is $N(10, 4)$ and are packed at random into boxes consisting of 25 chocolates.

Find the probability that

(a) a chocolate chosen at random weighs between 9.5 g and 10.5 g,
(b) the contents of a box weighs between 247 g and 253 g,
(c) the average weight of the chocolates in the box lies between 9.9 g and 10.1 g.

(a) Since W is $N(10, 4)$

$$P(9.5 < W < 10.5) = P(-0.25 < Z < 0.25)$$
$$= 2(\Phi(0.25) - \tfrac{1}{2})$$
$$= 0.197$$

(b) Let $T = W_1 + W_2 + W_3 + \ldots + W_{25}$
    where W_i represents the weights of the individual chocolates in the box.
    So T is $N(250, 100)$
    Thus

$$P(247 < T < 253) = P(-0.3 < Z < 0.3)$$
$$= 2(\Phi(0.3) - \tfrac{1}{2})$$
$$= 0.236.$$

(c) Let $\bar{W}$ represent the average weight of the chocolates in the box, so $\bar{W}$ is $N(10, 4/25)$. Thus

$$P(9.9 < \bar{W} < 10.1) = P(-0.25 < Z < 0.25)$$
$$= 2(\Phi(0.25) - \tfrac{1}{2})$$
$$= 0.197$$

Example 12.7
Two friends, John and Peter, take an average of 30 minutes to arrive at their place of work, each with a standard deviation of 5 minutes. If they travel independently find the probability that
(a) John arrives first,
(b) Peter arrives at least 2 minutes before John.

Let J and P be random variables representing the times for John and Peter respectively to arrive at work. So J is $N(30, 25)$ and P is independently $N(30, 25)$.

Consider the variate $D = J - P$; clearly D is $N(0, 50)$.

(a) $P(\text{John arrives first}) = P(D < 0)$
$$= 0.5.$$

(b) $P(\text{Peter arrives at least two minutes before John})$

$$= P(D \geqslant 2)$$
$$= P\left(Z \geqslant \frac{2}{\sqrt{50}}\right)$$
$$= P(Z \geqslant 0.283)$$
$$= 1 - \Phi(0.283)$$
$$= 0.389.$$

Exercise 12.1

1. X and Y are independent random variates, X is $N(6, 16)$ and Y is $N(3, 9)$. Calculate the mean and variance of Z when

(a) $Z = X + Y$
(b) $Z = X - 3Y$
(c) $Z = \bar{X}$,

where $\bar{X}$ represents the mean of random samples of size 4 taken from $N(6, 16)$,

(d) $Z = \bar{X} - \bar{Y}$,

where $\bar{X}$ represents random samples of size 9 taken from $N(6, 16)$ and $\bar{Y}$ represents random samples of size 16 taken from $N(3, 9)$,

    (i) In each case find the probability that Z is negative.

    (ii) In case (d), find $P(\bar{X} \leqslant \bar{Y})$.

2. The marks in a mathematics test are $N(10, 9)$. Find the probability that

    (a) 16 students chosen at random will have a mean mark less than 9,

    (b) 10 students chosen at random will have a mean greater than 11.

3. A particular type of resistor has a resistance, which is Normally distributed with a mean resistance of 200 ohms and standard deviation 10 ohms.

    If five resistors are connected in series find the probability that the resistance is between 980 ohms and 1020 ohms.

    A second type of resistor again with resistance normally distributed has a mean resistance of 100 ohms with standard deviation 4 ohms.

    If one of each type is connected in series find the probability that the total resistance will be in the range 300 ± 5 ohms.

4. If X is $N(3, 10)$ and $\bar{X}$ represents the mean of random samples of size 9. Find the value of k such that $P(\bar{X} < k) = 0.6$.

5. The variate X is related to the variates M and N by the formula

$$X = 3M - 2N$$

If M is distributed as $N(2, 2)$ and N is $N(4, 3)$, where M and N are independent, find the probability that

    (a) $M < N$

    (b) $X > 0$

6. Some toy cars are produced with weight normally distributed with mean 200 g and standard deviations 1 g. They are packed at random four to a box. Find the probability that

    (a) a box weighs more than 808 g, assuming the carton weighs 5 g,

    (b) a box weighs more than 808 g assuming the weight of the cartons is distributed as $N(6, 0.25)$.

7. An interview panel finds that the time to interview applicants for a job is approximately Normally distributed with mean 12 minutes and standard deviation 5 minutes. Find the probability that the panel can interview six applicants in less than 1 hour.

    If the first applicant is interviewed at 11 am find the probability that he will be able to catch a train leaving at noon if the journey to the station is normally distributed with a mean of 30 minutes and a standard deviation of 12 minutes.

8. The weights in kg of people using a certain lift are distributed as $N(70, 25)$. The lift has a maximum permissible load of 300 kg.

    (a) If four people are in the lift, find the probability that the maximum load is exceeded.

(b) If each person carries luggage whose weight is $N(10, 9)$, approximately, find the probability that the maximum load is exceeded if
   (i) 3 people enter,
   (ii) 4 people enter.

9. A man travels to work each day by walking to the station, waiting for a train and then travelling by train to his destination.

   Let X, Y and Z be the variates representing the time spent in minutes for each of the three stages of his journey with means and variances indicated below

	mean	variance
X	6	1
Y	8	3
Z	47	2

   (a) Calculate the mean and variance of the total time taken to travel to work.
   (b) If each of the variates is normally distributed, find the probability that he can complete the journey in under an hour.

10. Suppose beans are canned using cans with weight W g distributed $N(40, 1)$ and in such a way that the complete tin of beans has weight T g distributed $N(500, 10)$.
   (a) Describe the significance of the variate $B = T - W$ and write down its mean and variance.
   (b) The manufacturer wishes to state that the contents of the can weigh not less than x g, so that no more than 0.1% of the cans are underweight. Find the value of x to the nearest g.
   (c) If the manufacturer can control the mean weight of the beans supplied to each can, and wishes to state on the label that the contents weigh not less than 450 g with only 1% of the cans underweight, what should the setting be?

11.   A variate is $N(15, 4)$
   (i) Find the probability that
      (a) a single value drawn at random is less than 14,
      (b) a random sample of 9 items has a mean of less than 14.
   (ii) Find the size of a random sample required to ensure that the probability that the mean was less than 14 is not greater than 0.05.

12. The heights of boys in a certain school are approximately Normally distributed with mean 140 cm and standard deviation 10 cm. The heights of girls in the same school are approximately distributed as $N(135, 36)$.
   Two boys and two girls are selected at random.
   Find the probability that
   (a) the two boys differ in height by more than 6 cm,
   (b) the two girls differ in height by less than 6 cm,
   (c) the mean height of the two boys is less than the mean height of the two girls.

13. The height of a group of men is represented by the variate M cm which is $N(180, 100)$ and the height of their wives is represented by W which is $N(170, 144)$.

   (a) Calculate the probability that

       (i) a man chosen at random has a height less than 170 cm,

       (ii) a random sample of ten women has a mean height less than 169 cm.

   (b) Assuming that choice of partners is independent of height, find the probability that

       (i) a husband is taller than his wife,

       (ii) a wife is more than 2 cm taller than her husband,

       (iii) the mean height of a random sample of ten husbands differs by more than 2 cm from the mean height of their wives.

14. The independent random variables X_1, X_2, X_3 have means μ_1, μ_2, μ_3 and variances $\sigma_1^2, \sigma_2^2, \sigma_3^2$ respectively. State the mean and variance of the random variable $Y = a_1X_1 + a_2X_2 + a_3X_3$ where a_1, a_2, a_3 are constants.

   Norman Longlegs is a well known international athlete. His best event is the 'Hop-Step-Jump'. In this event, as the name suggests, an athlete takes a long run up to a starting board whereupon he hops, then steps and finally jumps. His recorded distance is from the starting board to his final position. Norman has observed that the three sections of his leap all follow independent Normal distributions. The Hop has a mean of 4 m and standard deviation 0.6 m; the Step has a mean of 2 m and standard deviation 0.5 m; the Jump has a mean of 3 m and standard deviation 0.5 m.

   The world record for this event is a distance of 10.5 m. What is the probability that Norman will break this record on any given attempt?

   In the European championships Norman is allowed three attempts. Assuming these attempts are independent, what is the probability that he breaks the world record at these championships?

   (You should assume that all leaps are fair and count.)

   AEB 1977

15. Two measurements X and Y are made on different subjects in a population, and are used in a medical diagnostic test based on $Z = X - 2Y$. In a particular population, X has a Normal distribution with mean 21.9 and standard deviation 3 while Y has an independent Normal distribution with mean 8.0 and standard deviation 2. Show that approximately 11.9% of the population will have a negative Z value.

   A random sample of 100 people is chosen from the population. Find the approximate probability that the sample contains more than 15 people that have a negative Z value.

   OXFORD

16. The discrete random variable R has the distribution

r	0	1	2	3
$P(R = r)$	0.6	0.2	0.1	0.1

Find the mean μ and variance σ^2 of the distribution.

Random samples of size 2 are chosen from the distribution. By considering all possible samples, find the probability distribution of the mean of the samples. State the mean and variance of this distribution in terms of μ and σ^2.

OXFORD

17. The weights of grade A oranges are Normally distributed with mean 200 g and standard deviation 12 g. Determine, correct to 2 significant figures, the probability that
    (i) a grade A orange weighs more than 190 g but less than 210 g,
    (ii) a sample of 4 grade A oranges weighs more than 820 g.

    The weights of grade B oranges are Normally distributed with mean 175 g and standard deviation 9 g. Determine, correct to 2 significant figures, the probability that
    (iii) a grade B orange weighs less than a grade A orange,
    (iv) a sample of 8 grade B oranges weighs more than a sample of 7 grade A oranges.

CAMB

18. The random variable X is Normally distributed with mean 1, variance 2. The random variable Y is also Normally distributed, independently of X, with mean 3, variance 4. $\bar{X}$ and $\bar{Y}$ denote the means of samples, each based on 5 observations, chosen at random from the corresponding distribution.
    Find
    (i) $E[(X - 2)^2]$,
    (ii) $E[(X + 3Y)^2]$,
    (iii) the distribution of $\bar{X} - \bar{Y}$,
    (iv) the probability that $\bar{X} > \bar{Y}$.

OXFORD

19. The variable X can take the values 0, 1, 2, 3, 4, 5, 6, 7, 8, 9 each with a probability of 0.1. Show that the mean of X is 4.5, and the variance of X is 8.25.

    Find the mean and variance of the distribution of the sum of 20 random digits chosen from 0 to 9 inclusive.

    Assuming this distribution to be approximately Normal, estimate the probability that the sum of 20 random digits lies in the range 80 to 109 inclusive.

AEB 1976

20. The random variables X_1 and X_2 are distributed independently with means μ_1 and μ_2, and variances σ_1^2 and σ_2^2 respectively. Write down the mean of $(X_1 + X_2)$, and prove that the variance of $(X_1 + X_2)$ is $(\sigma_1^2 + \sigma_2^2)$.

    The weight of a certain commodity sold each month by a shopkeeper is distributed Normally with mean 200 kg and standard deviation 15 kg. The weight sold in any month is independent of that sold in any other month. Find the probabilities that
    (i) in a given month more than 220 kg is sold;

(ii) in at least two of six consecutive months he sells less than 190 kg;

(iii) in a given two month period he sells more than 440 kg.

MEI

12.3 The Central Limit Theorem

The use of the term 'population' in statistics is due historically to the fact that statistics was applied mainly to the economic and sociological problems of societies. We now use the terms for any collection of objects, though there is usually a number associated with each object in this collection; the height of people or the score obtained from tossing a die, for instance.

We often model this situation by using a variate to represent the possible values associated with each member of the population. Thus we speak of the population having a probability density function (or in the discrete case, having a probability distribution) when we are actually referring to the associated variate. We often use the term *parent population* when referring to the population from which a sample was taken.

Suppose 50 random samples of size 10 are taken from a population having the discrete uniform distribution as indicated below.

x	0	1	2	3	4	5	6	7	8	9
p_r	$\frac{1}{10}$	$\frac{1}{10}$	$\frac{1}{10}$	$\frac{1}{10}$	$\frac{1}{10}$	$\frac{1}{10}$	$\frac{1}{10}$	$\frac{1}{10}$	$\frac{1}{10}$	$\frac{1}{10}$

For each sample of ten digits the mean can be computed. Thus if one sample was

$$2 \quad 0 \quad 0 \quad 7 \quad 7 \quad 7 \quad 8 \quad 4 \quad 6 \quad 3$$

then the mean $\bar{X}$ is 4.4.

The results obtained by author are recorded below.

$\bar{x}$	Frequency
2.0–2.9	3
3.0–3.9	12
4.0–4.9	19
5.0–5.9	14
6.0–6.9	2

As will be apparent from a histogram, this sample of means, $\bar{X}$, is bunched quite closely about the centre and one wonders what the shape of the distribution would be if the number of sample means obtained was increased indefinitely. Alternatively we might increase the size of the sample used to obtain each sample mean.

This tendency of $\bar{X}$, the mean of samples of size n (say) taken from a population, to bunch around the mean more than the original distribution was noted in the case of the Normal distribution.

Thus if X is $N(\mu, \sigma^2)$ then $\bar{X}$ is distributed as $N\left(\mu, \dfrac{\sigma^2}{n}\right)$

It is a remarkable fact that

Statement 12.2

If $X_1, X_2, \ldots, X_n$ are independent variates with identical distributions with mean μ and variance σ^2, and if C is a constant then

$$Y = CX_1 + CX_2 + \ldots + CX_n$$

tends to the Normal distribution as n tends to infinity.

What is remarkable about this result is that unlike the similar result in Section 12.1 we know nothing about the distributions of $X_1, X_2, \ldots, X_n$; other than the facts stated in the theorem.

In particular, if X is *any* variate with mean μ and variance σ^2, then $\bar{X}$, the distribution of sample means of size n, is approximately Normally distributed $N\left(\mu, \dfrac{\sigma^2}{n}\right)$ provided only that n is large enough. The approximation is usually satisfactory for $n \geqslant 30$ when X is reasonably symmetrical about the mean. The size of the sample n would have to be rather larger if X were particularly skew.

Example 12.8

Suppose X is the discrete variate with the probability distribution given below.

x	0	1
p_r	q	p

where $q = 1 - p$. It is easy to check that

$$E(X) = p$$

$$\text{Var}\ (X) = pq$$

Suppose further that a random sample $X_1, X_2, X_3, \ldots, X_n$ of size n is obtained, then the variate

$$Y = X_1 + X_2 + \ldots + X_n$$

is approximately $N(np, npq)$, provided n is large enough, say greater than 30. This follows as a direct application of the theorem taking $C = 1$. A moment's reflection will indicate that Y is in fact a discrete variate with a Binomial distribution, and thus the observation above is the justification that $B(n, p)$ can be approximated by $N(np, npq)$ (cf. Section 9.3).

12.4 Random samples

Populations come in all shapes and sizes and the random variables which model the populations can be continuous or discrete and take on an infinite or a finite number of values. A formal definition of a random sample of size n is as follows

Definition 12.1

> A random sample of size n taken from a parent population X is a set of random variables $X_1, X_2, \ldots, X_n$ such that each X_i has the same distribution as X and each pair of these random variables is independent.

The idea behind this definition is to ensure that at each stage each member of the population has an equal chance of being selected. This is often called *simple sampling*. Practical methods of obtaining a random sample can be found in Chapter 15.

A problem arises when we are dealing with a finite, and hence discrete, variate. In this case in order to attain simple sampling we must replace each item once selected so that it can (possibly) be selected again.

If we sample from a finite distribution without replacement then the probability of obtaining any particular value will depend on previous choices and so simple sampling does not apply.

In the example above, where samples of size 10 were selected from the digits, if the sampling had been without replacement, there would have been just one possible sample (if order did not matter) and thus the sampling distribution would contain just one item and so have zero standard error!

Statement 12.3

> If a sample of n values $X_1, X_2, \ldots, X_n$ is taken without replacement from a population X consisting of N values and at each stage each member of the available population is equally likely to be selected, then the statistic
>
> $$\bar{X} = \frac{X_1 + X_2 + \ldots + X_n}{n}$$
>
> is such that
>
> $$E(\bar{X}) = \mu$$
>
> $$\operatorname{Var}(\bar{X}) = \frac{\sigma^2}{n}\left(\frac{N-n}{N-1}\right),$$
>
> where X has mean μ and variance σ^2.

Note that if N is large compared with n then $\dfrac{N-n}{N-1} \simeq 1$ agreeing with our earlier results.

We shall not consider this problem further though we should bear in mind the need to check our assumptions carefully.

Exercise 12.2

1. Find the percentage of random samples of size 30 taken from a continuous population whose mean is 10 and with standard deviation 4 which will have means of 11 or more.

2. A sample of 100 items is taken from a continuous population with mean 10 and standard deviation 5.
   (a) Find the probability that the mean of the sample is more than 11 assuming the sample is chosen randomly.
   (b) It is stated that the mean of the sample is 8.2. What conclusions can you draw?

3. The mean score of a very large group of students taking an exam is 70 with standard deviation 8.

   Two random samples are taken, one of size 30 and the other of size 40. If $\bar{X}$ represents the mean mark of the first sample and $\bar{Y}$ the mean mark of the second, find the approximate distribution of

$$\bar{X} - \bar{Y}$$

   and justify your answer. Find the probability that the means of the two samples will differ by more than two marks.

4. If X and Y are continuous variates each with mean μ and variance σ^2, assuming standard notation explain what is meant by

$$\bar{X},$$
$$\bar{X} - \bar{Y},$$
$$X - Y,$$
$$\overline{X - Y}.$$

   In each case write down the mean and variance in terms of the size of the samples used.

5. The weight of a large collection of objects has a mean of 16 kg and variance of 5 kg. Find the probability that a random sample of 50 of these objects will have a mean weight between 15.5 kg and 16.5 kg.

6. A random sample of 1000 items was found to have a mean of 12.4. Find the probability that this sample could have been taken from a population whose mean is 12 with standard deviation 6.1.

7. One thousand ball-bearings have a mean weight of 4.9 g and standard deviation 0.3 g. Find the probability that a random sample of 60 ball-bearings will have a total weight of less than 292 g.

8. A Normal population has a mean of 60 and variance 30. What must be the size of a sample if its mean is to have at least a 90% chance of lying between 58 and 62?

9. The average score of students in an aptitude test is 72% with standard deviation 10%. What is the probability that two groups of students, consisting of 36 and 49 students respectively will differ in their mean scores by more than 3%?

10. The height of men is represented by the variate M with mean 180 cm and standard deviation 10 cm and the height of women is represented by the variate W with mean 170 cm and standard deviation 12 cm.

    Explain why it is not possible with the information given to find the probability that a man chosen at random has a height of less than 170 cm.

    Write down the mean and variance of

$$D = M - W$$

where M and W are independent.

    If a random sample of 30 marriage partners is chosen, and for each married couple the value of D is calculated and found to have a mean $\bar{D}$, explain why, assuming choice of marriage partner is independent of height, $\bar{D}$ can be approximated by a Normal distribution. Explain in words what $\bar{D}$ represents and find the probability that $\bar{D}$ is greater than 12.

11. A random sample taken from the continuous variate X measured to the nearest whole number has a frequency distribution

x	-1	-1	0	1	2	3
f	1	3	6	9	5	1

Estimate (a) $E(X)$, (b) Var (X), (c) the median of X, (d) the mean and standard error of the distribution of sample means of size 25.

12.5 Point estimates

The heights of 30-year-old males in England can reasonably be assumed to be approximately Normally distributed, but it is unlikely that we can guess the mean and variance of this distribution without further information. To find the mean and variance by measuring the height of each individual is prohibitively expensive and unnecessary.

Suppose we choose a random sample of 100 individuals from this population. We could certainly find the mean of this sample and would argue that this mean is quite a good estimate of the mean of the whole population. For if we imagine collecting means of many random samples of size 100 we would begin to build up the distribution of the sample means (of size 100). Now the standard error of this distribution is quite small (one tenth of the standard deviation of the original population) and we can thus be quite confident that our sample mean will be close to the mean of the distribution of sample means which, as we know, is equal to the mean of the original population. Thus we are confident that the mean of our sample is a good approximation to the mean of the population with the approximation improving as the size of the sample is increased since the standard error decreases.

Can we argue the same way when we try to estimate the variance of the population? We would be happy if we could use the variance of our sample as an estimate of the variance of the population. Again we may imagine variances being computed for various random samples to gradually build up the distribution of the sample variances. We have no information, as yet, concerning the shape of the distribution of sample variances, though we shall calculate its mean. Now if we can assume that the variance of the distribution of sample variances diminishes as the size of the random samples is increased, then we can be confident that the variance of our random sample is a good estimate of the mean of the distribution of sample variances (with the approximation improving as the sample is increased). So the variance of our random sample will be a good approximation to the population variance if the mean of the distribution of sample variance is a good approximation to the population variance.

This argument can be generalized as follows

Definition 12.2

> Suppose $X_1, X_2, \ldots, X_n$ is a random sample then a *statistic* is any function of these variables. Thus once the random sample is known the numerical value of the statistic is known.

For example, the following are statistics

$$\bar{X} = \frac{X_1 + X_2 + \ldots + X_n}{n}, \quad \text{the sample mean,}$$

$$S^2 = \frac{\Sigma(X - \bar{X})^2}{n} \quad \text{the sample variance.}$$

So also are the median, the range and the standard deviation.

Each statistic has a distribution called the sampling distribution of that statistic. Thus we have the sampling distribution of the mean and the sampling distribution of the variance. In each case the standard deviation is called the standard error of that statistic. We must be careful to distinguish between a statistic (which is a random variable) and any corresponding population parameter which is a (possibly unknown) constant.

Population parameter	Corresponding Statistic	
μ	$\bar{X}$	
σ^2	$\dfrac{\Sigma(X - \bar{X})^2}{n}$	(though there are other alternatives)
π	p	(proportions, cf. Section 14.1)

If we wish to use the values of a statistic R as estimates of a population parameter ρ (mean, variance, range, median etc.) then we speak of the statistic R as an *estimator* of the parameter ρ.

Definition 12.3

> The statistic R is said to be an *unbiased* estimator of the population parameter ρ if $E(R) = \rho$. That is, the mean of the sampling distribution of R is equal to the (usually) unknown population parameter ρ. In addition the statistic R is said to be *consistent* if $\text{Var}(R) \to 0$ as n (the size of the sample) $\to \infty$. That is, the variance of the sampling distribution diminishes as the sample size increases.

We shall be happy to use a value of R as an estimate of ρ if R is both unbiased and consistent. Though there are many other properties we would like an estimator to possess we shall be content with the two conditions above.

Statement 12.4

> The mean of a sample is an unbiased and consistent estimator of the population mean.

Suppose $\bar{X}$ represents the mean of a sample of size n taken from X with mean μ and variance σ^2. $E(\bar{X}) = \mu$, so $\bar{X}$ is an unbiased estimator and since $\text{Var}(\bar{X}) = \sigma^2/n$, $\bar{X}$ is a consistent estimator of μ.

It should, however, be borne in mind that any particular value of $\bar{X}$ could be a long way from the mean μ, though the probability of its being so decreases as the size of the sample increases.

Statement 12.5

> The variance of a sample is *not* an unbiased estimator of the population variance.

It is natural to ask whether

$$S^2 = \frac{\Sigma(X - \bar{X})^2}{n},$$

the variance of the sample, is both an unbiased and consistent estimator of the variance of the population from which it was taken.

Now

$$E(S^2) = E\left(\frac{\Sigma(X - \bar{X})^2}{n}\right)$$

$$= \frac{1}{n} E(\Sigma(X^2 - 2X\bar{X} + \bar{X}^2))$$

$$= \frac{1}{n} E(\Sigma X^2 - 2\bar{X}\Sigma X + \Sigma\bar{X}^2),$$

since $\bar{X}$ is a constant with respect to Σ though not with respect to $E(\,\cdot\,)$,

$$= \frac{1}{n} E(\Sigma X^2 - 2n\bar{X}^2 + n\bar{X}^2)$$

$$= \frac{1}{n} E(\Sigma X^2 - n\bar{X}^2)$$

$$= \frac{1}{n} (\Sigma E(X^2) - nE(\bar{X}^2)),$$

interchanging Σ and E is allowed
since $E(X + Y) = E(X) + E(Y)$,

$$= \frac{1}{n}\left(n(\sigma^2 + \mu^2) - n\left(\frac{\sigma^2}{n} + \mu^2\right)\right),$$

since $\sigma^2 = E(X^2) - \mu^2$

and $\dfrac{\sigma^2}{n} = E(\bar{X}^2) - \mu^2$

$$= \frac{n-1}{n}\sigma^2.$$

So S^2, the variance of the sample, is not an unbiased estimator of the variance of the population.

We now define a new statistic

Definition 12.4

$$\boxed{s^2 = \frac{n}{n-1}S^2 = \frac{\Sigma(X - \bar{X})^2}{n-1}}$$

s^2 is an unbiased and consistent estimator of the population variance.

$$E(s^2) = E\left(\frac{n}{n-1}S^2\right)$$

$$= \frac{n}{n-1}E(S^2)$$

$$= \frac{n}{n-1}\left(\frac{n-1}{n}\sigma^2\right)$$

$$= \sigma^2.$$

Both s^2 and (hence) S^2 are consistent since

$$\text{Var}(s^2) = \frac{(n-1)E((X-\mu)^4) - (n-3)\sigma^4}{n(n-1)},$$

though we shall not attempt to prove this!

Statement 12.6

> To estimate the variance of the parent population by using a sample, proceed as if calculating the variance of the sample, but divide by $N - 1$ instead of N. We use s^2 to denote this statistic.

Naturally the use of s^2 rather than S^2 is only particularly important when n is small, since for large values of n, $(n - 1)/n \simeq 1$.

We shall assume that s is an acceptable estimator for the standard deviation of the parent population.

Example 12.9
X is a continuous variate with *known* mean μ and $X_1, X_2, \ldots, X_n$ is a random sample from X.
Show that

$$\frac{\Sigma(X - \mu)^2}{n}$$

is an unbiased estimator of the variance of X.

$$E\left(\frac{\Sigma(X - \mu)^2}{n}\right) = \frac{1}{n} E(\Sigma(X^2 - 2X\mu + \mu^2))$$

$$= \frac{1}{n} \Sigma(E(X^2 - 2X\mu + \mu^2))$$

$$= \frac{1}{n} \Sigma(E(X^2) - 2\mu E(X) + \mu^2)$$

$$= \frac{1}{n} \Sigma[\sigma^2 + \mu^2 - 2\mu^2 + \mu^2]$$

$$= \frac{1}{n} \cdot n\sigma^2$$

$$= \sigma^2$$

Notice it is the fact that μ is known and is thus a constant which simplifies the calculations and ensures that the statistic is unbiased.

Exercise 12.3

1. Suppose X is a continuous variate and $\bar{X}$ and $\bar{Y}$ are independent samples of size n_1, n_2 respectively.
   Show that

$$V = \frac{\bar{X} + \bar{Y}}{2} \quad \text{and} \quad W = \frac{n_1\bar{X} + n_2\bar{Y}}{n_1 + n_2}$$

   are both unbiased and consistent estimators of the mean of X.
   In a situation such as this we think of the best estimator as the one with the smaller variance. Show that W is preferable in this case.

2. An experiment was performed five times resulting in the following measurements of a rod.

$$1.010 \quad 1.012 \quad 1.008 \quad 1.013 \quad 1.011$$

Calculate unbiased estimates for the mean and variance of possible measurements and give an estimate for the standard error of your estimate of the mean.

3. X is a continuous variate with mean μ and variance σ^2. Estimates of σ_1^2, s_1^2 and s_2^2 are obtained from random samples of size n_1 and n_2 respectively. Show that

$$\frac{(n_1 - 1)s_1^2 + (n_2 - 1)s_2^2}{n_1 + n_2 - 2}$$

is an unbiased estimator of σ^2.

4. An experiment was performed to determine the density of sulphuric acid. One scientist measured the density five times and obtained

$$1.841 \quad 1.840 \quad 1.844 \quad 1.842 \quad 1.841$$

and another obtained the four results

$$1.843 \quad 1.842 \quad 1.844 \quad 1.841.$$

Estimate the mean and variance of the density of sulphuric acid using

(a) the first five readings,
(b) all the data.

5. A random sample of five values is taken from a population with mean μ and variance σ^2. If the values are

$$8.1 \quad 6.5 \quad 4.9 \quad 7.3 \quad 5.9$$

estimate

(a) μ,
(b) σ^2,
(c) the standard error of the estimate for μ.

6. A random sample of observations of X was found to have a mean of 64 and a variance of 6. Obtain unbiased estimates of the mean and variance of X if the sample size was

(a) 10,
(b) 20.

7. The following frequency distribution represents the weights in kg of a random sample of 100 adults.

Weight	40	45	50	55	60	65
Number	8	28	36	18	9	1

Estimate the mean and standard error of the distribution of the sample means of size 100.

8. The mean of 10 readings of a variable was 8.7 with standard deviation 0.3. The following additional readings are given.

$$8.6 \quad 8.5 \quad 8.8 \quad 8.7 \quad 8.9$$

Estimate the mean and standard deviation of the set of possible readings using all the data available.

9. A random sample of 100 rods from a production line were measured and found to have a mean length of 12.132 with standard deviation 0.11. If a further sample of 50 items is taken, find the probability that the mean of this sample will be between 12.12 and 12.14.

10. A variate X has mean μ and variance σ^2. Two independent observations X_1 and X_2 are made. Obtain the values of the constants a and b so that $aX_1 + bX_2$ is an unbiased estimator of μ with the smallest possible variance.

11. (a) If $X_1, X_2, \ldots, X_n$ is a random sample from the distribution

$$f(x) = \begin{cases} \dfrac{xe^{-x/\theta}}{\theta^2} & x \geqslant 0 \\ 0 & \text{otherwise} \end{cases} \Bigg\} \theta > 0$$

show that $T = \sum\limits_{i=1}^{n} X_i/2n$ is an unbiased and consistent estimator of θ.

(b) Suppose $X_1, X_2, \ldots, X_n$ is a random sample from a Normal distribution with means μ and variance one. What is the distribution of the sample mean $\bar{X}$?

Calculate the sample size required to ensure that the probability of $\bar{X}$ being within 0.2 of μ is at least 0.95.

AEB 1978

12.6 Confidence intervals

As we have discovered in the previous section we can use a random sample to give an estimate of a population parameter. These estimates are, of course, just numbers and are called *point estimates*.

It is often preferable to give an *interval estimate* within which we have a reasonable degree of confidence that the parameter will lie. To fix this idea suppose we have a variate X which is $N(\mu, \sigma^2)$ where the variance σ^2 is *known* and μ is *unknown*.

Suppose a randomly observed value of X is found to be x. We have drawn in Fig. 12.1 several possible diagrams to represent $N(\mu, \sigma^2)$ where σ^2 has remained fixed, but we have selected several possible values for μ.

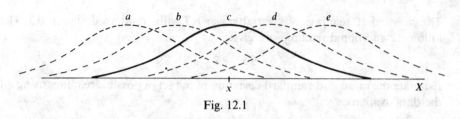

Fig. 12.1

Using the methods of the previous section, our point estimate would suggest c as the appropriate distribution to work with. It is, of course, possible that b or d is the correct distribution, though rather less likely that a or e is the correct one.

We can argue as follows

$$P(x \text{ is within one standard deviation of the mean})$$

$$= P(\mu - \sigma < x < \mu + \sigma)$$

$$= 0.68 \quad \text{(from Normal tables)}$$

that is, we are 68% confident that x is within one standard deviation of the (true) mean μ.

Similarly we can be

95% confident that x is within 1.96 standard deviations of the mean.

An entirely equivalent statement is that we can be

95% confident that the mean is within 1.96 standard deviations of x.

Example 12.10

Suppose we know that X is $N(\mu, 4)$ where the mean μ is unknown. If we obtain a value 3.4 as a random observation of X then we can be 95% confident that

$$3.4 - 1.96 \times 2 < \mu < 3.4 + 1.96 \times 2$$

that is

$$-0.52 < \mu < 7.32$$

The interval -0.52–7.32 is called a 95% confidence interval for μ and the numbers -0.52 and 7.32 are the 95% confidence limits for μ.

If a random sample of size 100 taken from the above population had a mean of $.1$, then working in the distribution of the sample means we can be 95% confident that

$$3.1 - 1.96 \times \frac{2}{10} < \mu < 3.1 + 1.96 \times \frac{2}{10}$$

that is

$$2.708 < \mu < 3.492$$

Thus, with the additional information, we can be as confident that μ lies in a much smaller interval.

The statement 'we are 95% confident that our confidence interval contains the mean' needs further clarification since either the mean is in the interval or it is not. The mean is a number not a variate or even a variable. What we intend by the statement is that in the long run this strategy for obtaining a confidence interval will be correct, in the sense that it contains the true and unknown mean 95% of the time.

We can select whatever degree of confidence we please, though there is a trade-off in that the more confidence we require the wider the confidence interval becomes or the more data we need if we wish to have a smaller interval.

Example 12.11

The standard deviation of the mass of a certain type of component is 0.2 g. How many components will have to be sampled in order to be 95% confident that the sample mean $\bar{X}$ will be within 0.05 g of the true mean?
The 95% confidence limits will be

$$\bar{X} \pm 1.96 \frac{\sigma}{\sqrt{n}}$$

assuming that either the mass of a component is normally distributed or that the sample size is large enough to invoke the Central Limit Theorem.
We require

$$1.96 \frac{\sigma}{\sqrt{n}} \leqslant 0.05$$

$$\Rightarrow \quad \frac{1.96 \times 0.2}{0.05} \leqslant \sqrt{n}$$

$$\Rightarrow \quad 7.84 \leqslant \sqrt{n}$$

$$\Rightarrow \quad 61.47 \leqslant n$$

and since n must be an integer the minimum sample size is 62 (large enough to be able to rely on the Central Limit Theorem to ensure that the distribution of sample means is normal which we required in order to use 1.96.)

We can use the idea of a confidence interval to give so called *interval estimates* of any population parameter (see Chapter 16 for a confidence interval for the variance) we can also find confidence intervals for distributions other than the Normal (see Section 14.2).

Confidence limits when the variance σ^2 is unknown

If we dealing with a large sample then we can use it to obtain our point estimate of the variance of the parent population and proceed as above. (Since the Central Limit Theorem ensures that the sampling distribution of the means is approximately Normal). If we are dealing with a small sample then the estimate of the variance is not accurate enough to use the above methods. If the sample is taken from a Normal distribution then we can solve the problem using the *Student t distribution* which is not dealt with in this book.

Exercise 12.4

1. A value of X taken from $N(\mu, 2)$ was found to be 0.9. Find a

   (a) 95%,
   (b) 90%,

   confidence interval for μ.

2. A sample of 16 values of X taken from $N(\mu, 3)$ was found to have a mean $\bar{X} = 4.2$. Write down the distribution of $\bar{X}$ giving its mean and standard error. Hence, find a 95% confidence interval for μ.

3. A random sample of 100 values of X was obtained with the following results

   $$\Sigma x = 960,$$

   $$\Sigma x^2 = 14\,625.$$

   Estimate the mean and variance of the population from which the sample was obtained.

   Describe the pdf of $\bar{X}$ explaining your reasoning, the means of samples of size 100, and hence obtain a 95% confidence interval for the mean of X.

4. A variate X is known to be $N(\mu, 25)$. A value of X is selected at random and found to be 6.8. Calculate a 99% confidence interval of the mean μ.

   If a random sample of size 15 had been found to have a mean of 6.8 then using all the information available find a new 99% confidence interval for the mean.

5. A random sample of 60 values of the variate X was found to have a mean of 5.9 and a variance 1.44. Find unbiased estimates of the mean and variance of X and hence find a 95% confidence interval for the mean.

6. A variate X has a variance of 4. How large must a random sample be in order to
   (a) ensure that the standard error is less than 0.05,
   (b) be 95% confident that the true mean is known to within 0.2 units?

7. A sample of 200 similar packets of breakfast cereal was examined, and the mass of the contents in each packet was noted. The following results were obtained: sample mean = 341.2 g, sample standard deviation = 0.92 g. Calculate a 95% confidence interval for the mean mass of contents in all packets of this type, stating any assumptions that you make.

   CAMB

8. From a population of mean μ and variance σ^2, a random sample of size n is taken. Assuming that the population is effectively infinite, show that the mean and variance of the sample mean are μ and (σ^2/n) respectively.

   In a large consignment of crates of tomatoes, the number of damaged tomatoes per crate has a Poisson distribution with mean λ. The mean number of damaged tomatoes per crate in a random sample of 60 crates is 22.7. Assuming that the sample mean has a Normal distribution, find approximate 95% confidence limits for λ.

Estimate the size of sample required for the difference between the confidence limits to be less than 1.

MEI

9. A continuous variate X follows the uniform distribution between 0 and a. That is its pdf is

$$f(x) = \begin{cases} \dfrac{1}{a} & 0 \leqslant x \leqslant a, \\ 0 & \text{otherwise.} \end{cases}$$

A single observation x is obtained. Use this to obtain an unbiased estimate of a and also a 90% confidence interval for a.

Chapter 13   Significance tests

13.1 Introduction

Suppose that the variate X is known to be Normally distributed with standard deviation 3 units where the mean μ, though unknown, is thought to be 10 units.

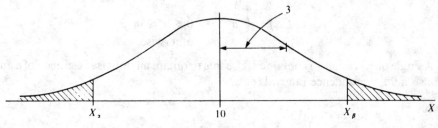

Fig. 13.1

The suggestion that $\mu = 10$ is simply an hypothesis, that is, a guess based perhaps on past experience or theoretical considerations. It is the usual practice to accept an hypothesis until events prove otherwise, so if a random observation of X was found to be 11 units (say) then we would be quite content with the assumption that $\mu = 10$. On the other hand if the observation turned out to be as high as 19 units, we would be placed in an awkward position. *Either the hypothesis is correct and a very rare event has occurred or the hypothesis is incorrect and should be replaced by an alternative.* In this case we would prefer to accept an alternative such as $\mu \neq 10, \mu > 10, \mu = 19$ or $\mu = 18$, since a mean of 10 is not only well outside the 95% confidence interval based on the observed value of 19, it is also outside the 99% confidence interval!

If the observation lies between 11 and 19 then the closer it is to 10, the happier we are to keep the original hypothesis, and the closer to 19 the more confident we are to reject the hypothesis and accept an alternative.

We see that it would be convenient to have a particular value, called the *critical value*, in mind. If the observation is greater than this critical value we reject otherwise we do not.

To fix these ideas suppose we take as our original hypothesis, called the *null hypothesis*, NH, that $\mu = 10$ and this time as our *alternative hypothesis* AH, that $\mu \neq 10$.

In this case if a very large or very small value is observed then we would prefer to reject the NH in favour of the AH, otherwise we shall continue to accept the NH. Our problem is simply to choose the critical values X_α and X_β in Fig. 13.1 to enable us to make a decision. *The shaded region can be interpreted as the probability that we reject the NH, when it is in fact true!*

We can control this probability by altering our choice of critical values X_α and X_β. If we move them further away from the mean of the distribution we are working with, thus reducing the shaded area, then we reduce the probability of rejecting the NH, when it may, in fact, be false. Similarly if we increase the shaded region, by moving the critical values closer to the mean, then we make it easier to reject the NH even though it may well be true.

In practice statisticians have found that making this shaded area called the *rejection region* equal to 5% of the total is about the right balance. This will be referred to as testing at the 5% level. We shall see later that making this rejection region equal to 1% or even 0.1% is sometimes more appropriate.

A moment's thought and a glance at the Normal distribution tables suggests that the critical values X_α and X_β should each be 1.96 standard deviations from the mean.

We now have our decision procedure. We establish the number of standard deviations our observed value x of X, is from the mean $\mu = 10$.
That is compute:

$$z = \frac{x - 10}{3}, \text{ called the } test\ statistic.$$

$$z = \frac{x - \mu}{\sigma}$$

Now if $-1.96 \leqslant z \leqslant 1.96$ that is $|z| \leqslant 1.96$ we do not reject the null hypothesis, and if $z < -1.96$ or $z > 1.96$, that is $|z| > 1.96$, then we reject the null hypothesis in favour of the alternative hypothesis. In such a case we say that a *significant result* has occurred.

Example 13.1

The variable X is distributed as $N(\mu, 9)$, where it is thought that $\mu = 10$. Test this hypothesis against the alternative that $\mu \neq 10$ given that a random sample of 16 observations was found to have a mean of 12.1.

We shall consider

$$\text{NH}: \mu = 10$$

$$\text{AH}: \mu \neq 10$$

On the basis of the NH, which we assume to be true until proven otherwise, we know that

$$X \text{ is } N(10, 9)$$

and thus $\bar{X}$, the distribution of the means of samples of size 16, is $N\left(10, \frac{9}{16}\right)$

Working in this distribution, the value of our test statistic is

$$z = \frac{12.1 - 10}{\frac{3}{4}} = 2.8$$

and, as before, the critical values are -1.96 and 1.96.

Since $2.8 > 1.96$ we reject the NH in favour of the AH. The evidence supports the view that $\mu \neq 10$.

Example 13.2

The length of screws produced by a particular machine is known to be Normally distributed with variance 0.16 mm^2. The machine was set to produce screws with a mean length of 20 mm, but it is now thought that the screws recently produced are too long. A random sample of 25 screws was found to have a mean length of 20.1 mm. Test at the 5% level whether the mean length of the screws produced by the machine has increased.

The first point to note is that if the sample had produced a mean length of less than 20 mm then there would be no evidence that the mean length had increased! It is only if the mean length of the sample is significantly larger than 20 mm that we would wish to accept that the machine was incorrectly set.

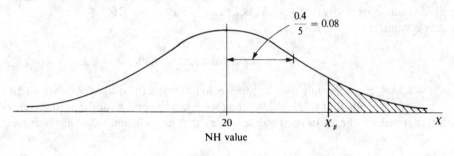

Fig. 13.2

A careful reading of the question suggests that we consider.

$$NH : \mu = 20$$

$$AH : \mu > 20$$

Now on the basis of the NH the variate X representing the length of screws produced by the machine is $N(20, 0.16)$ so $\bar{X}$, the means of sample of size 25, is $N\left(20, \dfrac{0.16}{25}\right)$

Our test statistic is

$$z = \frac{20.1 - 20}{\dfrac{0.4}{5}} = 1.25$$

We have mentioned previously that only large positive values of the test statistic would persuade us to reject the NH in favour of the AH. So the rejection region is only on one side of the mean (Fig. 13.2). In this case there is only one critical value, 1.645 standard deviations above the mean, to ensure that the shaded area is still 5% of the total.

Now $1.25 < 1.645$ so we do not reject the NH and so we have no evidence that the machine is incorrectly set.

In the example above we have used a so-called one-tailed test, since the rejection region is confined to one side of the mean. This is typical when the alternative hypothesis is of the forms $\mu > k$ or $\mu < k$ for some constant k. In the earlier example we were dealing with a two-tailed test, which arises when the AH is of the form $\mu \neq k$.

13.2 Some definitions associated with tests of significance

The null hypothesis NH
The null hypothesis is a statement which is accepted until the evidence proves otherwise. It will usually translate into a statement about the mean of the variate used in the *model*, it usually suggests that the mean takes a specific value which we shall refer to as the NH value.

The alternative hypothesis AH
This statement usually takes one of two basic forms:

 (i) $\mu \neq$ NH value;
(ii) $\mu <$ NH value (or $\mu >$ NH value).

In case (i) we use a *two-tailed test* and in case (ii) a *one-tailed test*.
 If a *significant result* occurs then the NH is rejected in favour of the AH.

The significance level
The significance level is usually taken to be 5% (though sometimes we consider 1% or 0.1% levels). This level determines the *critical values* of the test. The critical values also depend on whether we are dealing with a *one-tailed* or *two-tailed test* and on the *model*.
 We shall restrict ourselves to considering the Normal distribution or a discrete distribution though there are many other important distributions which occur in practice.

The critical values
If the test statistic falls outside the critical values, then a significant result has occurred and the NH is rejected in favour of the AH.
 The critical values given in the following table relate to a Normal distribution.

level of significance	*one-tailed*	*two-tailed*
5%	1.645	± 1.96
1%	2.33	± 2.575
0.1%	3.1	± 3.3

The one-tailed or two-tailed test
A *one-tailed test* has a rejection region on just one side of the mean, but a *two-tailed test* has rejection regions on both sides.

The test statistic
In a test of significance we are dealing with a *model* which depends to some extent on the NH. The test statistic is the number of standard deviations the observed value is from the mean. Its particular form depends on the model.

Significant results

If the *test statistic* falls in the rejection region then we say that a significant result has occurred when testing at the 5% level. When testing at the 1% level we speak of a highly significant result and at the 0.1% level a very highly significant result.

The rejection region

The rejection region is determined by the critical values. If the *test statistic* falls in the rejection region then we can reject the NH in favour of the AH.

The model

There are only certain situations that we can cope with at this stage.

Model 1

If a single value of the Normal variate X is observed where X has variance σ^2, then the model to use is N(NH mean, σ^2).

Model 2

A random sample of X is used where either
(a) X is Normally distributed with known variance σ^2. Then the model to work with is N(NH mean, σ^2/n) where n is the size of the sample.
(b) A *large* random sample is obtained so we again can work with N(NH mean, σ^2/n), where n is the size of the sample and σ^2 is either known or estimated from the sample.

Model 3

The number of 'successes' is observed from a Binomial situation $B(n, p)$ and if n is large enough we can use $N(np, npq)$.

Example 13.3

A coin was tossed 100 times and 41 heads were observed. Using the Normal approximation to the Binomial distribution test at the 5% level (of significance) whether the result provides evidence that the coin is

(a) biased,
(b) biased in favour of tails,
(c) biased in favour of heads.

When a coin is tossed 100 times the number of heads obtained, X, follows the Binomial distribution $B(100, p)$, where p represents the probability of obtaining a head on a single toss of a coin.

It is sensible to assume that $p = 1/2$ until events prove otherwise, so we shall adopt this as our null hypothesis in each case.

(a)   NH $: p = \frac{1}{2}$

AH $: p \neq \frac{1}{2} \Rightarrow$ two-tailed test

On the assumption of the NH we can assume that the variate X is approximately distributed as

$$N(50, 25)$$

Since the number of heads obtained was 41 the test statistic is

$$z = \frac{41 - 50}{5}$$

$$= -1.8.$$

(We do not take the correction factor into account.)

We are testing at the 5% level with a two-tailed test so the critical values are ± 1.96: and since

$$-1.96 \leqslant -1.8 \leqslant 1.96,$$

we do not reject the null hypothesis.

(b)   NH $: p = \frac{1}{2}$

AH $: p < \frac{1}{2}$

reflecting the bias in favour of tails. In this case we have a one-tailed test. Again we are working with the approximation $N(50, 25)$. Only a low value of X will make us wish to reject the NH so the critical value is -1.64. Our test statistic remains the same at

$$z = -1.8$$

and since

$-1.8 < -1.64$ a significant result has occurred and we can reject the null hypothesis in favour of the alternative. That is the evidence supports the view that the coin is biased in favour of tails.

(c)   NH $: p = \frac{1}{2}$

AH $: p > \frac{1}{2}$

There is no point in taking the analysis further since we need a large number of heads to obtain a significant result.

It is important to realise however that the choice of alternative hypothesis depends on the nature of the problem not on the particular test statistic which is observed.

Example 13.4
A factory produces components with a mean life of 100 h (hours). After a change in the production process a random sample of 100 such components was found to have a mean life of 102.5 h with a variance of 99 h^2. Test whether the new process has improved the life of the components now produced.

We take X to be the variate representing the life of the components with mean μ. As usual we take the conservative view on the null hypothesis. Namely

$$NH : \mu = 100 \quad \text{(there has been no change)}.$$

Since we are looking for an improvement we take

$$AH : \mu > 100 \quad (\Rightarrow \text{a one-tailed test}).$$

We work with the distribution of sample means of size 100, which by the Central Limit Theorem will be approximately Normally distributed. We do not know the variance of X, but we can estimate it from the sample as $\dfrac{100}{99} \cdot 99 = 100 \, h^2$ to give an unbiased estimate.

Since X is approximately $N(100, 100/100)$ our test statistic is

$$z = \frac{102.5 - 100}{1} = 2.5$$

We shall as usual test at the 5% level and so the critical value is 1.64, since we are dealing with a one-tailed test.

Now $2.5 > 1.64$ so a significant result has occurred and we can reject the NH in favour of the AH. That is the change in the production process has improved the mean life of the components.

Exercise 13.1

1. It has been established over a long period of time in a particular factory that the net weight of the contents of cans of beans is normally distributed with mean 510 g and standard deviation 4.1 g.

   A recently filled can was found to have a content weighing 498 g. Assuming the standard deviation has not altered test at the 5% level whether the filling process is faulty.

2. A variate X is Normally distributed with standard deviation 17.6 units. A value of X was observed and found to be 151.9. It is thought that the mean of X is either 120 or perhaps a little more. On the basis of the evidence what can you say? Is there any evidence to support the view that the mean is not 120?

3. A person claims that he can tell the difference between margarine and butter. In a test he correctly distinguishes between the butter and margarine 25 times out of forty trials. Are the results significant at the 5% level?

4. A Normal population has variance 0.5 unit^2 but an unknown mean. Five values observed at random were found to be

$$3.6 \quad 4.5 \quad 1.9 \quad 6.2 \quad 3.8$$

Test at the 5% level that the mean is 5.

5. A factory produced electric light bulbs with a mean life which was thought to be 2000 h and a standard deviation known to be 100 h. A random sample of 36 light bulbs are found to have a mean life of 2021 h. Does this sample support the view that the mean life of the bulbs produced by the factory has changed.

6. A pilot wishes to check whether the average depth of a section of the river mouth is still 6.1 fathoms as his chart suggests. He takes 30 soundings at random points in the region and finds the mean depth to be 5.8 fathoms with a variance of 1.2 fathoms^2. Test at the 5% level whether the mean depth has changed.

7. A random sample of size 20 was selected from a Normal variate with variance 25 units^2 and found to have a mean of 20 units. It has been assumed that the mean of the variate was 22 units. Test this assumption at the 5% level against both the AH that the mean is less than 22 and the AH that the mean is not 22.

8. A sample of 25 school children took an intelligence test for which it is known that the marks are distributed as $N(100, 225)$ for all children of a certain age. The mean of the sample was found to be 105.2. Does this provide evidence that the children were not selected at random?

9. The weights of men in a certain town were found to have a mean of 64 kg with standard deviation 8 kg. A sample of 100 citizens was selected at random and found to have a mean weight of 61 kg. How could you explain those results?

10. A certain component has been found to have a mean life of 100 h with variance 100 h^2. After a change in the production process, a sample of 100 such components selected at random was found to have a mean life of 102 h. Test whether the new process has had a beneficial effect on the lifetimes of these components assuming that the variance has remained unaltered.

11. A sample of 78 students took an entrance examination. The mean score was 68% with a standard deviation of 10%. Long experience has suggested that the mean would be about 70%. Does this evidence suggest that the 78 students are worse than usual?

12. The times taken for a salesman to travel between two shops on 50 occasions were found to have a mean of 1 h 4 m with variance 40 m^2. He claims that this provides evidence that the journey is taking longer these days since he changed to a smaller car, as his mean journey time used to be only 1 h. Can you agree with the salesman?

13. A random sample of size 40 is found to have a mean of 5.6 units and variance 0.8 units^2. Test at the 5% level if the sample could have come from a population with mean 6 units.

14. A random sample of 21 potatoes from a large crop had mean mass 0.12 kg with standard deviation of 0.02 kg. Give estimates for the mean and variance of the whole crop.

What would be the expected mean and variance of a random sample of 11 potatoes from the same crop?

Discuss briefly whether a mean of 0.13 kg for another sample of 40 potatoes from the crop would indicate that the sample was not random.

AEB 1978

13.3 Type I and type II errors

In significance testing we are concerned with whether to accept or reject a null hypothesis NH; which we hope is closely related to whether the null hypothesis is true or not.

There four possibilities as indicated in Fig. 13.3.

	Accept NH	Reject NH		Found innocent	Found guilty
NH True	Correct decision	Incorrect decision Type I error	In fact innocent	Correct	Incorrect
NH False	Incorrect decision Type II error	Correct decision	In fact guilty	Incorrect	Correct

Fig. 13.3   A table to show the logical possibilities in significance testing: compare with the situation found in a court of law.

Definition 13.1

A type I error occurs when the null hypothesis is rejected when it is in fact true.

A type II error occurs when the null hypothesis is accepted when it is in fact false.

The probability of making a type I error is equal to the area of the shaded region in the diagram Fig. 13.4, since it is precisely when the test statistic falls into this region that the alternative hypothesis is accepted. It is interesting then that the probability of making a type I error equals the level of significance of the test, and so is under our direct control.

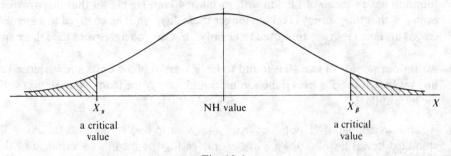

Fig. 13.4

The probability of a type II error depends on which alternative hypothesis is actually true.

Suppose, for example, that the null hypothesis is false and the correct hypothesis is as indicated in Fig. 13.5.

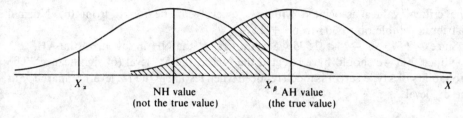

X_α NH value (not the true value) X_β AH value (the true value) X

Fig. 13.5

To make a type II error we have to accept the NH when it is the AH which is correct. The probability of making this error is equal to the shaded area in Fig. 13.5.

After a moment's thought one can see that in order to reduce the probability of a type II error the critical values X_α and X_β (we are assuming a two-tailed test) need to move closer to the NH value. But this *increase* in the significance level increases the probability of a type I error! Conversely reducing the probability of a type I error, by decreasing the level of significance, means that the probability of a type II error is increased. Of course the actual probability of a type II error depends on which AH is actually true so we cannot take the analysis further in this book.

Finding the appropriate balance between the two types of errors depends on which error produces the most serious consequences.

Using the analogous situation in a court of law (see Fig. 13.3) a type I error corresponds to finding an innocent man guilty; and a type II error corresponds to finding a guilty man innocent. In this country we aim to reduce the probability of a type I error and in consequence accept an increased value for the type II error.

13.4 Other significance levels

Example 13.5

A random sample of 64 items has a mean of 6.5 units. The variance of the parent population is 16 units^2. Test at the 1% level (of significance) the hypothesis that $\mu = 7.5$ units.

Let X be the variate representing the parent population. We know that the variance of X is 16 units^2 and will assume that the mean is 7.5 units for our test. There is no indication what the alternative hypothesis is to be so we shall adopt the following:

$$NH : \mu = 7.5 \text{ units},$$

$$AH : \mu \neq 7.5 \text{ units} \quad (\Rightarrow \text{two-tailed test}).$$

On the basis of the Central Limit Theorem and the NH we can assume that $\bar{X}$, the means of samples of size 64 is $N(7.5, 16/64)$.

The test statistic is

$$z = \frac{6.5 - 7.5}{\frac{1}{2}} = -2$$

The critical values associated with a 1% level can be found from the Normal distribution table on p. 303 to be ± 2.575.

Since $-2.575 < -2 < 2.575$ we cannot reject the NH in favour of the AH.

Notice that we should have rejected the NH at the 5% level (of significance). The probability of a type I error is much reduced when testing at the 1% level compared with the 5% level.

13.5 The difference of two means

Example 13.6

During an epidemic of Rubella the ages of patients treated at a particular hospital were collected with the following results.

	Number of patients	Mean age	Variance
Men	154	13.3	86.5
Women	129	14.1	74.3

Is there a significant difference between the mean ages of men and women who contract the disease?

None of the models mentioned in Section 13.2 is applicable to this situation.

Let the variate M represent the age of men contracting the disease and W represent the age of women contracting the disease. Setting $E(M) = \mu_M$ and $E(W) = \mu_W$ then the question really is whether $\mu_M = \mu_W$ or not. The problem is that we know nothing about the distribution of M or W nor do we know the variance of either variate.

We can estimate the variance of M and W by using the variance of the samples given:

$$s_M^2 = \frac{154}{153} \times 86.5 \simeq 87.1,$$

$$s_W^2 = \frac{129}{128} \times 74.3 \simeq 74.9.$$

In practice when dealing with such large samples the variance of the sample is often used as the estimate of variance of the parent population.

Though we know nothing of the distribution of M and W we can claim that $\bar{M}$, the distribution of sample means of size 154 and $\bar{W}$ the distribution of sample means of size 129 are approximately Normally distributed. Thus the difference $\bar{M} - \bar{W}$ will also be Normally distributed.

We now start our test and consider:

$$\text{NH} : \mu_M = \mu_W,$$

$$\text{AH} : \mu_M \neq \mu_W \quad (\Rightarrow \text{two-tailed test}).$$

The model we shall work with is

$$\bar{M} - \bar{W},$$

where

$$E(\bar{M} - \bar{W}) = E(\bar{M}) - E(\bar{W})$$

$$= E(M) - E(W)$$

$$= \mu_M - \mu_W$$

$$= 0, \qquad \text{by the null hypothesis,}$$

$$\text{Var} (\bar{M} - \bar{W}) = \text{Var} (\bar{M}) + \text{Var} (\bar{W}) \quad \text{assuming indepen-}$$
$$\text{dent samples}$$

$$= \frac{\text{Var } M}{154} + \frac{\text{Var } W}{129}$$

$$\simeq \frac{87.1}{154} + \frac{74.9}{129}$$

$$\simeq 1.15.$$

So the standard deviation of $\bar{M} - \bar{W}$ is 1.07 (approximately). Our test statistic is

$$Z = \frac{-0.8 - 0}{1.07}$$

$$= -0.75,$$

which is clearly not going to give us a significant result at the 5% level. So there is no evidence to suppose that there is any difference between the mean ages of men and women who contract Rubella.

Exercise 13.2

1. The following data gives the weights of a sample of 1 year old children.

	Sample Size	Mean Weight	s^2
Boys	64	9.2	0.8
Girls	81	8.8	1.2

Test at the 5% level the hypothesis that there is no difference between the average weight of one-year-old boys and girls.

2. It is decided to check whether the average speed of cars leaving a city is different from those entering the city around noon on a particular day. Vehicles were timed

over a 1 mile stretch of road with the following results.

	Leaving	Entering
Number of vehicles	50	50
Mean time to cover. 1 mile in seconds	17.03	18.28
Variance	8.85	9.2

Determine whether the difference between the means is significant at the 5% level.

3. A sample of size 9 is taken from a Normal distribution with variance 4 and another sample of size 12 is obtained from a Normal distribution with variance 1. If the means of the samples are 24.5 and 25.2 respectively, is there any evidence to suggest that there is no difference between the means of the population from which the samples were drawn?

4. Two similar machines make screws. From the first a random sample of 120 screws is taken and from the second a sample of 150 screws is taken. The results obtained from measuring the lengths of the screws is as follows:

	First machine	Second machine
Sample mean	4.1 cm	4.6 cm
Sample variance	0.1 cm^2	0.5 cm^2

Do you think the machines were set to give different lengths of screws?

5. A Bank Manager wished to know the effect of a recent increase in the interest rates on the amount held in 'current' accounts. The following data was collected.

	Sample size	Mean of sample £	s
Month before increase	60	134	25
Month after increase	60	128	32

Write a short report for the manager giving your interpretation of the results in 'plain' English.

6. The 58 children entering School A and the 62 entering School B were each given the same Maths examination. After one year they were given a further common examination.
   The following results were obtained.

| | Upon entry | | After one year | |
	mean	standard deviation	mean	standard deviation
School A	53%	10%	51%	7%
School B	50%	10%	48%	7%

Show that there was no significant difference upon entry, but there was after 1 year.

7. The mean number of days absence for 150 children in the first year of a particular school was 3.1 days per child and in the fourth year it was 3.6 days per child for the same number of students. The estimates for the variance of the number of days absence are 3.2 days^2 and 6.5 days^2 respectively. Examine whether there is evidence that children tend to take more days off in the fourth year than in the first year.

8. Explain briefly what is meant by a significance level of $\alpha\%$.

A group of 200 workers in the 'public sector' compared their wages with a group of 300 performing a similar job in the 'private sector'.

Use the data given below to test whether the wages in the 'public sector' are lower than those in the 'private sector' at the 5% level.

Number in sample	mean weekly wage (£)	standard deviation (£)
200	96	20
300	100	15

Find the smallest value of α for which a significant result is obtained using the above data.

9. The heights of men can be assumed to be Normally distributed with standard deviation 0.11 m.

In 1928 the mean height of men in a certain city was 1.72 m. In a survey in 1978 the mean height of a random sample of 16 men from the same city was 1.77 m. On the hypothesis that the population mean height has not changed, calculate the probability of obtaining a sample mean height greater than that measured.

In another survey in 1978 the mean height of a random sample of 32 men from a second city was 1.73 m. Assuming that the population mean heights are the same in the two cities, calculate the probability that a difference in sample mean heights greater than that measured would be obtained.

MEI

10. Two independent random variables X and Y are distributed with the same means and with standard deviations σ_x and σ_y respectively. The mean of N_x independent readings of X is $\bar{X}$, and the mean of N_y independent readings of Y is $\bar{Y}$. Show that $(\bar{X} - \bar{Y})$ has zero mean and standard deviation equal to

$$\left(\frac{\sigma_x^2}{N_x} + \frac{\sigma_y^2}{N_y} \right)^{1/2}$$

The viscosity of a certain liquid is measured by two standard methods A and B. It is known that the standard deviations of measurements by these methods are 0.13 units and 0.32 units respectively. The mean of 10 measurements by method A is 16.18 units and the mean of 50 measurements by method B is 16.44 units. Do you consider that the difference between these means is significant? (You may assume that any relevent variable is distributed Normally.)

MEI

11. A company has two factories which produce a certain material. Their daily outputs are independent and are normally distributed with means 12 and 15 tonnes and standard deviations 0.5 and 0.6 tonnes respectively.

Some changes are made in the staffing of the plants and it is thought that the means may be affected but the standard deviation will not. A check over 100 days reveals mean production figures of 11.92 and 15.10 tonnes respectively. Test at the 5% level to see if either of these is significantly different from its established mean daily value.

Test also whether or not the difference of the means is significantly different from the value which would have been expected.

Calculate a symmetrical two-sided 95% confidence interval for the difference of the means based on these 100 day samples of output figures.

JMB

13.6 A word of warning

It is appropriate to give a word of warning at this stage. We know that if we take small random samples from a Normal distribution with variance σ^2 and mean μ then the sampling distribution of $\bar{X}$ is itself Normal with mean μ and variance σ^2/n.

In a test of significance using $\bar{X}$ as a model involves the test statistic

$$z = \frac{\bar{x} - \mu}{\dfrac{\sigma}{\sqrt{n}}},$$

where n is the size of the sample and μ is the NH value of the mean. Of course, z is a particular value of the standardised variate Z which is $N(0, 1)$.

However, if the variance of the parent population is unknown and has to be estimated from a small sample then the statistic

$$t = \frac{\bar{x} - \mu}{\dfrac{s}{\sqrt{n}}},$$

is no longer Normally distributed. This statistic follows the so-called Student t distribution which is not dealt with in this book.

13.7 Small sample tests using the Binomial distribution

Example 13.7
A boy claims he can throw a head with a coin four times out of six on average.

Calculate the probability that he will throw 5 or more heads in 6 throws:
(a) if the coin is unbiased;
(b) if his claim is justified.

He is invited to throw the coin six times, his claim being accepted if he throws five or six sixes. Find the probability of his claim being accepted when in fact he has no particular skills.

(a) Let the variate X represent the number of heads obtained, then X follows $B(6, 1/2)$.
Thus

$$P(X \geqslant 5) = P(X = 5) + P(X = 6)$$

$$= 6 \left(\frac{1}{2}\right)^5 \frac{1}{2} + \left(\frac{1}{2}\right)^6$$

$$= \frac{7}{64}$$

$$\simeq 0.109$$

(b) In this case X is $B(6, 2/3)$. So

$$P(X \geqslant 5) = P(X = 5) + P(X = 6)$$

$$= 6 \left(\frac{2}{3}\right)^5 \frac{1}{3} + \left(\frac{2}{3}\right)^6$$

$$= \frac{256}{729}$$

$$\simeq 0.351.$$

If he has no particular skills the appropriate model to use is $B(6, 1/2)$. In this case the probability that his claim is accepted is $\simeq 0.109$.

Exercise 13.3

1. Eight people are asked to choose the heaviest rock from three of similar weight. If six or more choose correctly would you agree that people are able to distinguish the heaviest rock? Explain your argument with some care.

2. A boy claims that he is able to throw an unbiased coin and obtain more heads than tails in the long run.
    Find the probability of his obtaining 8 or 9 heads in 9 tosses of the coin if (a) the coin falls at random, (b) if in fact he obtains twice as many heads as tails in the long run.
    He is challenged about his ability and is invited to toss a coin nine times. His claim will be accepted if he obtains 8 or 9 heads. Find (i) the probability that his claim is accepted when he has no special ability, (ii) the probability that his claim is not accepted when he can in fact throw twice as many heads as tails in the long run.

3. A coin has probability p of falling heads when tossed. Write down, in terms of p, expressions, which need not be simplified, for
    (i) the probability that if the coin is tossed four times all four tosses will fall alike (that is, all heads or all tails),
    (ii) the probability that if the coin is tossed seven times six or more of the tosses will fall alike.

The following two procedures have been suggested for testing whether the coin is unbiased.

Procedure 1: Toss the coin four times and conclude that it is biased if all four tosses fall alike.

Procedure 2: Toss the coin seven times and conclude that it is biased if six or more of the tosses fall alike.

Show that the two procedures are equally likely to lead to the conclusion that the coin is biased when in fact it is unbiased.

Determine which of the two procedures is the less likely to lead to the conclusion that the coin is unbiased when in fact the probability of a head in any toss is equal to 2/3.

JMB

4. A manufactured item has probability p of being less than 20 mm long. In a random sample of 10 of these items, let X denote the number of items whose lengths are less than 20 mm. Write down an expression, which need not be simplified, for $P(X = r)$, where r is an integer from 0 to 10 inclusive.

   The observed value of X is to be used as a basis for testing the hypothesis that m, the mean length of the manufactured items, is equal to 20 mm against the alternative hypothesis that m is less than 20 mm. The decision rule is to reject the hypothesis $m = 20$ mm only if $X \geqslant k$ where k is an integer.

   It is known that the lengths of the manufactured items are Normally distributed with a standard deviation of 0.1 mm whatever the value of m.

   (a) Find the value of p when $m = 20$ mm, and hence find the smallest value of k for the significance level of this decision rule to be less than 0.05.

   (b) Find the value of p when $m = 19.8$ mm. If $k = 10$, calculate the probability, to 2 decimal places, that the decision will be to accept the hypothesis $m = 20$ mm when in fact the true value of m is equal to 19.8 mm.

JMB

5. A man, Mr Knowall, claims to be able to recognise a well known brand, A, of whisky blindfolded. On 8 occasions he is given two glasses in random order, one containing brand A and the other containing brand X. On 7 occasions he names the brands correctly. Would you consider his claim justified?

   The same experiment is repeated the following day, and this time he names the brand correctly in 6 out of 8 trials. How does this affect your opinion of his claim?

AEB 1977

6. A member of an organoleptic (tasting) panel claims to be able, by drinking the resulting beverage, to determine whether tea or milk was poured first into a cup.

   (a) If she diagnoses correctly r in a series of n such trials, explain how you would decide whether to accept her claim or whether she is just guessing. (Do not consider any approximate methods in this part.)

   (b) Would you accept her claim if she diagnosed 8 out of a total of 10 correctly?

   (c) Would you accept her claim if she diagnosed 60 out of a total of 100 correctly?

OXFORD

Chapter 14 The distribution of proportions and percentages

14.1 Introduction

We are often interested in the proportion or the percentage of a population which possesses a common property or attribute. The percentage voting for a particular party in a General Election or the proportion of defective items produced by a particular machine for example.

Suppose π is the actual proportion of a population possessing a particular property. If a random sample of size n is observed then R, the number of items possessing the property, will follow the Binomial distribution providing the usual conditions of a Bernoulli trial are met.

In this case the mean of R will be $n\pi$ and the variance of R will be $n\pi(1 - \pi)$. The variate $P = R/n$ represents the proportion of the sample possessing the property. We have chosen to use π as the true proportion rather than p so as not to confuse with a particular value of P, the variate representing the proportions in samples of size n. Now

$$E(P) = E\left(\frac{R}{n}\right)$$

$$= \frac{1}{n} E(R)$$

$$= \pi$$

$$\text{Var}(P) = \text{Var}\left(\frac{R}{n}\right)$$

$$= \frac{1}{n^2} \text{Var}(R)$$

$$= \frac{n\pi(1 - \pi)}{n^2}$$

$$= \frac{\pi(1 - \pi)}{n}.$$

Thus any particular value p of P will be an unbiased and consistent estimator of π.

If n is large we know already that R can be approximated by the Normal distribution $N(n\pi, n\pi(1 - \pi))$ provided π is neither too large nor too small. In such a case P will be approximately $N(\pi, \pi(1 - \pi)/n)$.

Example 14.1
Forty people out of a random sample of 100 could tell the difference between butter and

margarine in an experiment. Find the probability that in a further sample 50 or more can tell the difference.

The actual proportion π of the population which can distinguish correctly is unknown, but an unbiased and consistent estimate is $40/100 = 0.4$.

Since $n = 100$ is large we can approximate the distribution of P, the proportion in samples of size 100 which can tell the difference, by

$$N\left(0.4, \ \frac{0.4 \times 0.6}{100}\right)$$

that is

$$N(0.4, \ 0.0024).$$

We have to establish

$$P(P \geqslant 0.5) = P\left(Z \geqslant \frac{0.1}{\sqrt{0.0024}}\right)$$

$$= 1 - \Phi(2.04)$$

$$\simeq 0.021.$$

The alert reader will notice that the above question could be solved using only the Binomial distribution without recourse to the distribution of proportions. With practice it will be found that the above method is often the most convenient. He will also recall that a continuity correction was made when using the Normal approximation to the Binomial distribution. When considering the distribution of proportions the correction factor is of the form $1/2n$ which may be small enough to ignore if n is large.

14.2 Confidence intervals for proportions

If the variate P represents the proportions of samples of size n possessing a particular property or attribute and n is large enough, then we have stated that P is $N[\pi, \pi(1 - \pi)/n]$, where π is the actual proportion possessing the property.

If the proportion p possessing the property is obtained from a random sample of size n then we could obtain a 95% confidence interval for π as

$$p \pm 1.96 \sqrt{\text{Var}(P)},$$

if n is large enough to use the normal approximation. The problem of course is that the Var (P) involves π, which is presumably unknown. If we estimate the value of π as p then we have as our 95% confidence interval approximately

$$p \pm 1.96 \sqrt{\frac{p(1 - p)}{n}}.$$

In such circumstances we often use

$$p \pm 2 \sqrt{\frac{p(1 - p)}{n}}.$$

Example 14.2

A random sample of 200 voters was canvassed and 64 said they would vote for the Conservative candidate. Find an approximate 95% confidence interval for the true proportion. It was decided in view of the importance of this election that a further sample was to be canvassed to ensure that the proportion would be known to within 10% of the true proportion with 95% confidence. Find the size of the sample required in order to ensure this.

Our unbiased estimate of π is $p = 64/200 = 0.32$. So an approximate 95% confidence interval for π is

$$0.32 \pm 2\sqrt{\frac{0.32 \times 0.68}{200}},$$

that is 0.25–0.39.

An approximate 95% confidence interval for π is of the form

$$p \pm 2\sqrt{\frac{p(1 - p)}{n}}.$$

To guarantee that π is known to within 10% we want

$$2\sqrt{\frac{p(1 - p)}{n}} \leqslant \frac{10}{100}p$$

$$\Rightarrow \frac{4p(1 - p)}{n} \leqslant \frac{p^2}{100}$$

$$\Rightarrow n \geqslant \frac{400(1 - p)}{p}$$

$$\Rightarrow n \geqslant \frac{400 \times 0.68}{0.32}$$

$$\Rightarrow n \geqslant 850.$$

Exercise 14.1

1. If the probability of a male birth is 0.51, find the probability that there will be more girls than boys in a random sample of

   (a) 5,   (b) 100 children.

2. A set of 30 'mixed' ability children was set a test and 20 passed.
   Find (a) an estimate of the proportion of the whole year group who would pass; (b) the standard error of this estimate; (c) the probability that less than 15 pass the test when another 'mixed' ability group of 30 takes the test.

3. If C represents the *percentage* of a random sample of size n possessing a particular property for which the true *proportion* is π, find the mean and variance of C and discuss the conditions for which C will be approximately Normally distributed.

4. In a random sample of 200 industrial accidents it was found that 112 were due to unsafe practices. Find 95% confidence limits for the true

   (a) proportion,   (b) percentage.

5. Find the maximum possible value of the expression $p(1 - p)$.

   Find the size of a sample required to estimate an unknown proportion to within a maximum error of 0.05 with at least a 95% degree of confidence if (a) no information is available about the true proportion and so the maximum standard error is used, (b) π is estimated to be 0.2.

6. In an inspection of 4000 electrical components built to a very high specification just 10 were found to be defective. Find an upper 95% confidence limit for the true proportion of defectives in the production of these components.

7. A chemical manufacturing company wishes to erect a factory in an urban area. For economic considerations two thirds of the population are in favour of the factory, and, for environmental reasons, one third are against it.

   To estimate these proportions the company consults a random sample of 100 people. Assuming that the number in favour of the factory in such a sample has a Binomial distribution, write down the mean and standard deviation of this distribution.

   Using the Normal distribution as an approximation to the Binomial distribution, find the probabilities that in the sample there are (i) more than 70, (ii) less than 60 in favour of a factory.

   In a sample of 200 people, the percentage of the sample in favour of the factory is obtained. Find the mean and standard deviation of this percentage.

   The company decides that if less than 60% of the sample are in favour it will not build the factory. Estimate the least sample size which is necessary for the probability of such a decision to be less than 0.2%.

                                                                                            JMB

14.3 Testing a proportion

The technique for testing the hypothesis that the true proportion π has some particular value is similar to previous tests. We consider the variate P representing the proportion possessing some particular property in samples of size n. Providing n is large enough P can be assumed to be approximately Normally distributed with mean π and variance $\pi(1 - \pi)/n$.

Example 14.3
According to Mendel's theory of inheritance certain crosses of peas should give yellow and green peas in the ratio 3:1. In an experiment 207 yellow and 73 green peas were obtained. Does this data support the theory?

We shall consider the variate P representing the proportion of yellow peas in samples of size 280 where the true proportion is π. We shall take

$$NH : \pi = \frac{3}{4}$$

$$AH : \pi \neq \frac{3}{4} \quad (\Rightarrow \text{two-tailed test}).$$

On the basis of the NH, P is approximately

$$N \left(\frac{3}{4}, \frac{\frac{3}{4} \cdot \frac{1}{4}}{280} \right), \quad \text{that is, } N(0.75, 0.00067).$$

The proportion of yellow peas in our sample is $207/280 \simeq 0.739$, so the test statistic is

$$z = \frac{0.739 - 0.75}{\sqrt{0.00067}} = -0.42.$$

Testing at the 5% level on a two-tailed test the critical values are ± 1.96, so clearly we do not have a significant result and there is no reason to doubt the theory of inheritance as propounded by Mendel on the basis of this evidence.

14.4 The difference of two proportions

Example 14.4

Two drugs, Getwel and Pikup are administered to two separate groups of patients. The first to 60 patients of whom 20 respond and the second to 70 of whom 25 respond. Is there any evidence of a difference in effectiveness of these drugs?

Let the variate P_G represent the proportion who respond to the first drug in samples of size 60 and P_P represent the proportion who respond to the second drug.

Setting the true proportions to be π_G, π_P respectively, then we shall test the following:

$$NH : \pi_G = \pi_P = \pi \quad (\text{say})$$

$$AH : \pi_G \neq \pi_P \quad (\Rightarrow \text{two-tailed test})$$

We do not know the value of π and so combine the data to obtain the following estimate

$$\pi \simeq \frac{20 + 25}{60 + 70}$$

$$= \frac{45}{130} \simeq \frac{9}{26}$$

Consider the variate $P_G - P_P$, we know that it will be approximately Normally distributed, since we are dealing with the difference of two independent variates each of which is approximately Normally distributed as, in each case, the sample size is large.

Now using the NH we have

$$E(P_G - P_P) = 0$$

$$\text{Var } (P_G - P_P) = \text{Var } (P_G) + \text{Var } (P_P)$$

$$= \frac{\pi(1 - \pi)}{60} + \frac{\pi(1 - \pi)}{70}$$

$$= \frac{9}{26} \cdot \frac{17}{26} \left(\frac{1}{60} + \frac{1}{70} \right)$$

$$\simeq 0.007,$$

So $P_G - P_P$ is approximately $N(0, 0.007)$.

Now the observed difference is

$$\frac{20}{60} - \frac{25}{70} = -0.024,$$

so the test statistic is

$$z = \frac{-0.024 - 0}{\sqrt{0.0007}}$$

$$= -2.8.$$

We are dealing with a two-tailed test so at the 5% level the critical values are ± 1.96. Since $z < -1.96$, a significant result is obtained. The evidence supports the view that there is a difference in effectiveness of the drugs.

Exercise 14.2

1. In the Example 14.4 test whether Pikup is significantly more effective than Getwel at the 1% level.

2. During the first part of this century it was found that the proportion of male births was 0.51. More recently it was found that of a random sample of 40 000 births 20 665 were boys. Would you agree with the suggestion that the sex ratio has changed?

3. A coin is tossed n times and 48% of the time it comes down heads. How large must n be before we can conclude at the 5% level that the coin is biased?

4. A die is tossed 100 times and only 10 sixes are obtained. Test at the 5% level whether the die is biased.

5. Write down your own estimate of the proportion of left-handed people in this country.

    In a random sample it is found that 35 are left-handed and 245 right-handed. Test at the 5% level whether the proportion is different from your guess. Establish the range of values you could have guessed and yet not obtain a significant result in the above test.

6. A garage maintains a stock of 25 cars and sold 10 cars per week on average, until VAT was raised to 15%. During the following week only 5 cars were sold. Would you agree that the increase of VAT had adversely affected the sales?

7. Until recently approximately 50 out of every 100 patients survived a particular disease. When a new drug was tried 81 out of 150 patients survived. Does this provide significant evidence that the drug is effective?

8. An organisation interviews a randomly chosen sample of 1000 adults from the population of the United Kingdom, and 517 of those interviewed claim to support the Conservative party. A second organisation independently interviews a random sample of 2000 adults of whom 983 claim to support the Conservative party.
   (a) Verify that the results of the two organisations do not differ significantly at the 5% level.
   (b) Obtain a symmetric 99% confidence interval (based on the combined results) for the proportion of the population who claim to support the Conservative party.

CAMB

9. Explain the basic ideas underlying significance tests, discussing, in particular the role played by the 'null hypothesis'. Illustrate your answer by reference to the following problem.

   The proportion of male to female lambs born in a certain breed of sheep is known, on the basis of very extensive records, to be 52 : 48. Given that 156 progeny of a ram (in his first season) consisted of 96 males and 60 females, is there significant evidence that the ram differs from the breed average in the sex ratio of his offspring?

   Explain, with reasons, whether you consider a one-sided or two-sided test to be appropriate to this problem.

JMB

10. The probability of success in each of a long series of n independent trials is constant and equal to p. Explain how 95% approximate confidence limits for p may be obtained.

   In an opinion poll carried out before a local election, 501 people out of a random sample of 925 declare that they will vote for a particular one of two candidates contesting the election. Find 95% confidence limits for the true proportion of all voters in favour of this candidate.

   Do you consider there is significant evidence that this candidate will win the election?

AEB 1977

Revision exercise C

1. In a sequence of random digits the number of non-zero digits between two successive zeros is counted and found to be N. Find the probability distribution of

N and hence find the mean value of N. Find also the probability that a particular value of N is less than the mean.

In a sequence of random digits 201 zeros were observed and of the resulting 200 values of N, 170 were less than the mean. Use the Normal approximation to the Binomial distribution to test whether the digits should be considered to be from random number tables.

2. The variate X has the uniform distribution, when X takes the values 0, 1, 2 or 3. Find
   (a) the mean and variance of X,
   (b) the mean and variance of $\bar{X}$, where $\bar{X}$ represents the mean of samples of size two taken at random with replacement from X.
   Derive the probability distribution of $\bar{X}$ directly and use it to obtain the mean and variance of $\bar{X}$.

3. Define type I and type II errors in testing hypotheses.
   A box is known to contain either (H_0) ten white counters and 90 black counters or (H_1) 50 white counters and 50 black counters. In order to test hypothesis H_0 against hypothesis H_1, four counters are drawn at random from the box, without replacement. If all four counters are black, H_0 is accepted. Otherwise it is rejected.
   Find the size of the type I and type II errors for this test.

                                                                    AEB 1976

4. State, giving your reasons, which probability distribution you would expect to serve as a model for describing:
   (i) the number of times a total score of 'six' is obtained when two dice are tossed ten times,
   (ii) the number of breakdowns of a particular machine in a period of one month,
   (iii) the precise weight of the contents of a bag of flour.

5. A random sample of the men marrying in a town had the following age distribution.

Age, in years	Under 18	18–24	25–31	32–38	39–45	46–52	Over 52	Total
Coded value		0	1	2	3	4		
Frequency	0	101	62	25	10	2	0	200

By using the suggested coding, or otherwise, estimate the mean and variance of the population from which the sample is drawn. (Quote your results to 2 DP.)

The mean age of men at marriage in the large country in which the town is situated is 25.90 years. Test whether the men in the town differ significantly from the men in the whole country in the age at which they marry. (State your null hypothesis and the significance level you use.)

                                                                    OXFORD

6. In an association football match x is the number of goals scored by the home team and y is the visiting team's score. The frequency distribution for the total score, $x + y$, for first division matches in 1972–3 is as shown.

Total score	0	1	2	3	4	5	6	7	8
Frequency	35	82	138	99	61	32	7	7	1

Calculate the mean and standard deviation of the distribution.

Draw a frequency polygon for the data.

It is known that the mean value of x for these games is 1.56; calculate the mean value of y.

Assuming that x and y are independent variables, indicate how you would calculate the variance of y if that of x were known. Give a reason why the assumption of independence may not be justified in this case.

JMB

7. Two species of insect, type A and type B, are distinguished by measuring a particular characteristic x. The probability distribution of x for insects of type A is

$$p_A(x)dx = C_1(1 - x)^2 dx \quad (0 \leqslant x \leqslant 1)$$

and the probability distribution of x for insects of type B is

$$p_B(x)dx = C_2 x^2(1 - x)dx \quad (0 \leqslant x \leqslant 1).$$

Find the values of the constants C_1 and C_2 and calculate the mean and the standard deviation of each distribution.

It is decided that, when x has been measured for a particular insect, then if $x \leqslant \frac{1}{2}$ the insect will be classified as type A and if $x > \frac{1}{2}$ the insect will be classified as type B. Find

(a) the probability that an insect of type A is wrongly classified as type B,

(b) the probability that an insect of type B is wrongly classified as type A.

MEI

8. Two samples of sizes M and N have means $\bar{x}_M$ and $\bar{x}_N$ respectively, and standard deviations s_M and s_N respectively. The two samples are combined to form a sample of size $(M + N)$ whose mean and standard deviation are $\bar{x}$ and s respectively. Prove that

$$(M + N)\bar{x} = M\bar{x}_M + N\bar{x}_N, (M + N)s^2 = Ms_M^2 + Ns_N^2 + \left(\frac{MN}{M + N}\right)(\bar{x}_M - \bar{x}_N)^2.$$

[For a sample of size n whose members are $x_1, x_2, \ldots, x_n$, the mean $\bar{x}$ and the standard deviation s are defined by

$$n\bar{x} = \Sigma x \quad \text{and} \quad n^2 s^2 = n\Sigma(x^2) - (\Sigma x)^2.]$$

An examination is taken by 100 students of whom 90 are in centre A and 10 in centre B. The marks of the 90 students in centre A have a mean of 49.7 and a standard deviation of 12.3. The marks $x_1, x_2, \ldots, x_{10}$ of the 10 students in centre B

satisfy $\Sigma x = 520, \Sigma(x^2) = 32\,020$. Calculate the mean and standard deviation of the marks of all 100 students.

MEI

9. The random variable X can take all values between 0 and a inclusive, where $a > 0$. Its probability density function $f(x)$ is zero for $x < 0$ and $x > a$, and, for $0 \leqslant x \leqslant a$, satisfies

$$f(x) = (A/a)\exp(-x/a),$$

where A is a positive constant. Show by integration that $A = 1.582$ to 3 decimal places.

   Also use integration to find to 2 decimal places
   (i) the probability that X is less than $\frac{1}{2}a$;
   (ii) the number λ for which there is a probability $\frac{1}{2}$ that X is less than λa.

MEI

10. A box contains nine numbered balls. Three balls are numbered 3, four balls are numbered 4 and two balls are numbered 5.

   Each trial of an experiment consists of drawing two balls without replacement and recording the sum of the numbers on them, which is denoted by X. Show that the probability that $X = 10$ is 1/36, and find the probabilities of all other possible values of X.

   Use your results to show that the mean of X is 70/9, and find the standard deviation of X.

   Two trials are made. (The two balls in the first trial are replaced in the box before the second trial.) Find the probability that the second value of X is greater than or equal to the first value of X.

MEI

11. Explain the meaning of the term confidence interval.

   The expressions $\Sigma(x_i - \bar{x})^2/n$ and $\Sigma(x_i - \bar{x})^2/(n - 1)$ are both used in connection with variance for a set of observations $x_1, x_2, \ldots x_n$. Explain the circumstances in which each is used and hence distinguish between them.

   For such a set of 81 independent observations from a Normal distribution, $\Sigma x_i = -36$ and $\Sigma x_i^2 = 736$. Construct a 95% confidence interval for the mean of their probability distribution, and use it to test whether this mean is likely to be zero.

   Explain, either by carrying out the appropriate calculations, or otherwise, how you would conduct this test without using a confidence interval.

OXFORD

12. Define the Binomial distribution, stating clearly the conditions under which it applies, and derive its mean.

   A man canvasses people to join an organisation. For each new recruit he receives 25p. The probability that a person he canvasses will join is 0.2. Calculate to three decimal places the probability that he will obtain three or more recruits from ten people canvassed. State the amount of money that he would be expected to obtain on average from ten canvassings. State how many people he would need to canvass each evening to average ten new recruits per evening.

Calculate the number of people he must arrange to canvass to be 99% certain of obtaining at least one recruit.

JMB

13. Explain the fallacy in each of the following arguments, and give the correct conclusions.

(a) Y is a Normal random variable with mean 0, variance 25. The probability that $Y \leqslant 5$ is obtained from the table of the Normal integral with $x = (5 + \frac{1}{2})/5$, so we conclude that this probability is 0.8643.

(b) X_1 and X_2 are independent random variables with means μ_1 and μ_2, variances σ_1^2 and σ_2^2 respectively. Hence $X_1 - X_2$ has mean $\mu_1 - \mu_2$, variance $\sigma_1^2 + \sigma_2^2$, and $(X_1 - X_2) + X_2$ has mean $(\mu_1 - \mu_2) + \mu_2 = \mu_1$, variance $(\sigma_1^2 + \sigma_2^2) + \sigma_2^2$ $= \sigma_1^2 + 2\sigma_2^2$. But $(X_1 - X_2) + X_2 = X_1$, so we conclude that X_1 has mean μ_1, variance $\sigma_1^2 + 2\sigma_2^2$.

(c) Two independent observations, 26 and 45, arise from Normal distribution of unknown (possibly the same) means and the same variance, 25. Hence the mean of the first distribution must lie between 16 and 36 and that of the second between 35 and 55. These ranges overlap, so we conclude that with 95% certainty the two distributions are the same.

OXFORD

14. In observations of a particular type of event, the probability of a positive result of any one observation is independent of the results of other observations and has the value θ, the same for all observations. In n observations the proportion giving positive results is p. State the mean and standard deviation of the probability distribution of p. Say also how and in what circumstances this probability distribution can be approximated by a Normal distribution. Show that, according to this approximation, the probability that p satisfies the inequality

$$|p - \theta| < 1.96 \sqrt{\frac{\theta(1 - \theta)}{n}} \text{ is } 95\%.$$

In a set of 100 observations of this type, 90 gave a positive result. Obtain an inequality of the above form, and by squaring both sides of the inequality calculate from the roots of a quadratic equation an approximate 95% symmetric confidence interval for the value of θ for the type of event observed.

JMB

15. The random variables $X_1, X_2, \ldots, X_n$ are independent and each has a Normal distribution with mean μ and variance 1. The random variable $\bar{X}$ is defined to be $(X_1 + X_2 + \ldots + X_n)/n$. Determine, in terms of n, the value v which is such that, when $\mu = 0$, the probability of $\bar{X}$ exceeding v is 0.05.

For this value of v it is desired that the probability of X being less than v when $\mu = 0.2$ should be at most 0.10. Calculate the smallest possible value of n which satisfies this requirement.

CAMB

16. A shopkeeper's fortnightly sales of a commodity can be regarded as having a Normal distribution with mean 200 kg and variance 225 kg^2. Find the probability that the sales of the commodity are less than 185 kg. When he reorders, delivery takes two weeks. Determine to the nearest kilogram the stock which he should have in hand when he reorders so that the probability that he will not run out before the new delivery arrives is 95%.

Five shops combine for bulk buying purposes and their fortnightly sales of the product are independent and have Normal distributions with means 200, 240, 180, 260 and 320 kg and variances 225, 240, 225, 265 and 270 kg^2 respectively. Write down the mean and variances of their total fortnightly demand and determine to three significant figures the total level of stock at which they must reorder so that the probability of not running out before the new delivery is 99%.

Find the probability that the total amount sold by the shops in ten weeks is more than 6200 kg.

JMB

17. The score, S, gained by an expert rifleman with a single shot, is a random variable with the following probability distribution:

$$P(S = 8) = 0.01, \ P(S = 9) = 0.29, \ P(S = 10) = 0.70.$$

(a) Use a Normal approximation to determine the probability that the rifleman obtains six or more scores of 8 in a series of 900 independent shots.

(b) Find the expectation and variance of S.

(c) Use a Normal approximation to determine the probability that the rifleman scores less than 96 with ten independent shots.

CAMB

18. Define the Poisson distribution and derive its mean and variance.

In the first year of the life of a certain type of machine, the number of times a maintenance engineer is required has a Poisson distribution with mean four. Find the probability that more than four calls are necessary.

The first call is free of charge and subsequent calls cost £20 each. Find the mean cost of maintenance in the first year.

JMB

19. A farmer has an orchard containing large numbers of two varieties (A and B) of apple trees. One year the farmer selected at random one tree of each variety, and kept a careful count of the fates of all the apples from these two trees. Some apples fell from the trees before picking time; some were eaten by insects; some eaten apples fell and some apples remained uneaten and on the trees until picking time. The farmer's results are given below.

	Variety A			Variety B	
	Fallen	On tree		Fallen	On tree
Eaten	150	50	Eaten	90	50
Uneaten	40	160	Uneaten	20	240

You may assume, when answering the questions below, that the fate of an individual apple was independent of the fate of all other apples.

(a) Before any apple had fallen or been eaten, the farmer selected at random a variety A apple and stated that it would not fall before picking time. Estimate the probability that he was correct.

(b) At picking time the farmer accidentally trod on a fallen apple. Assuming that this apple was equally likely to have been any one of the fallen apples, estimate the probability that it was of variety A.

(c) Give an approximate symmetric 95% confidence interval for p_A, the proportion of variety A apples remaining on the tree and uneaten until picking time.

(d) The proportion of variety B apples remaining on the tree and uneaten until picking time is p_B. Determine whether there is evidence at the 0.1% significance level of a difference between p_A and p_B.

CAMB

20. A player throws a die whose faces are numbered 1 to 6 inclusive. If the player obtains a six he throws the die a second time, and in this case his score is the sum of 6 and the second number; otherwise his score is the number obtained. The player has no more than two throws.

Let X be the random variable denoting the player's score. Write down the probability distribution of X, and determine the mean of X.

Show that the probability that the sum of two successive scores is 8 or more is 17/36. Determine the probability that the first of two successive scores is 7 or more, given that their sum is 8 or more.

CAMB

21. A bag contains three red and seven white balls. Five balls are selected and the number R of red balls noted before the balls are returned. Find the probability distribution of R and hence find the mean and variance of R.

Chapter 15 Sampling techniques

15.1 Methods of sampling

One way to acquire knowledge of a population is to investigate every member. A prime example is the *population census* conducted every 10 years in the United Kingdom. Until quite recently it was usual to study every member of a population. With the increase in data this procedure tended to defeat its own purpose, in that by the time the data had been analysed it was quite often out of date. In any case it can be prohibitively expensive if carried out to excess, and in some circumstances the very investigation destroys the subject, as for example, when testing missiles by firing them.

We have discovered enough to realise that a sample from a population can be used to answer questions of the whole population with, if not complete accuracy, within a sufficiently accurate range to be of value to the investigator and at much less expense. The assumption is that such a sample can be taken to represent the population and this is achieved by considering a *random* sample which is large enough to produce sufficiently accurate results in further analysis.

As is intuitively clear a random sample is obtained by ensuring that each member of the population has an equal chance of being chosen *cf.* Section 12.4. In this way probability theory can be used to show that it is unlikely that any bias is introduced, which would prevent the sample from being representative, and that as the sample size is increased we can be even more confident that this is so. Our immediate concern is to investigate methods of selecting a sample.

Random sampling

If the population to be studied is small then each member can be assigned a number, and correspondingly numbered strips of paper placed in a hat. After mixing well they can be taken out one by one (with or without replacement depending on the nature of the population) to obtain a random sample of the required size.

If the population is sufficiently large to make the above procedure tedious then *random number tables* can be used instead (cf. p. 304). Such tables could have been compiled using the process described above, by placing the digits $0, 1, \ldots, 9$ on strips of paper and then choosing digits with replacement, though in practice a speedier method is used. However, these tables can be used to draw a sample from quite large populations with relative ease.

Example 15.1

Explain how to obtain a random sample of size 10 from a population whose individuals have been assigned the numbers

$$(a)\ 0\text{–}99\,999,$$

$$(b)\ 1\text{–}330.$$

(a) An inspection of the random number tables shows that, as an aid in reading, the numbers are arranged in groups of five digits ranging from 00 000 to 99 999. Now taking 00 000 as 0 and 00 001 is 1 and so on, each group of 5 digits can be considered as a possible member of our sample. Further, since they are printed at random any consecutive sequence of 10 such groups could be used to define our sample. It is wiser to select consecutive groups since an apparently arbitrary selection could result in an unintentional bias. For this reason it is important that the choice of the first member of the sample should be selected with great care to ensure, as far as is possible, a 'random' starting point and not rely on personal preferences.

(b) In this case the same idea can be adopted rejecting any number larger than 330. To reduce wastage one might decide to obtain a suitable number by ignoring the first two digits in each group and then subtract a suitable multiple of 330 from numbers which are still too large and ignore numbers 991 or above. So for example

Random number			Sample number
60 637	→ 637	→	307
39 772	→ 772	→	112

Stratified sampling
In theory we could choose a sample of size 2000 from the electorate in the United Kingdom, using random number tables. In practice this would be very time consuming, not only in choosing the sample, but in trying to contact the corresponding electors. One way of dealing with this problem is to divide the population into separate categories such as the constituencies and choose some of these at random taking account of the difference in size. Then in each chosen constituency a sample of polling districts could be randomly selected, with the names chosen at random from each of the appropriate electoral registers. If a large number of names was required from a list such as the electoral register, then it is common to choose a name at random and then to select every tenth or hundredth name. A major advantage of the use of stratified sampling is that it can improve the accuracy of the results.

Quota sampling
In practice even the above procedure can be too time consuming or expensive. An alternative strategy is to choose a sample so that it reflects as far as possible the population to be studied. If the population is classified into groups in various ways, say by sex, age and occupation then an investigator may be told to ensure that for each classification the proportion in his sample is similar to that of the whole population. The *choice* of which person to interview is then left to him.

15.2 Simulation
Suppose we wish to obtain a sample of scores obtained from an unbiased die. One method is to shake such a die and record the scores. A possible objection, of course, is that the die we choose can only be said to be an approximation to an (ideal) unbiased die. An alternative method is to use random number tables to obtain suitable values. In this way we have simulated the experiment of tossing a die.

On the other hand, suppose we wish to obtain a sample of scores from a die which is biased, so that $P(6) = 1/2$ and $P(1) = P(2) = P(3) = P(4) = P(5) = 1/10$. It would be very difficult to obtain a die with the appropriate properties. The use of random number tables is very convenient in this case. We would probably interpret the (random) digits as

Random digit	0	1	2	3	4	5	6	7	8	9
Valve of variate	6	1	2	3	4	5	6	6	6	6

We can extend this idea to sampling from a continuous variate for any degree of accuracy we require.

Example 15.2
Show how to obtain a random sample from the continuous range 3–7 including the end points so that each member of the sample is given to two decimal places.

The method employed here is to consider consecutive triples from random number tables and translate as follows:

$$356 \rightarrow 3.56$$

$$614 \rightarrow 6.14$$

$$712 \rightarrow$$

(we would reject this value since it is outside the range of interest).

We can in fact sample from any given distribution as the following example indicates.

Example 15.3
Show how random number tables can be used to obtain a random sample of values from the variates X and Y distributed as follows:

(a) x	0	1	2	3	4
p_r	$\frac{1}{10}$	$\frac{1}{10}$	$\frac{3}{10}$	$\frac{4}{10}$	$\frac{1}{10}$
(b) y	0	1	2	3	
p_r	$\frac{7}{40}$	$\frac{11}{40}$	$\frac{9}{40}$	$\frac{13}{40}$	

(a) In order to obtain a random sample we need to ensure that each member of the population is equally likely to be selected and yet still reflect the fact that we are three times as likely to obtain a 2 as a zero. We might decide to allocate random digits as follows.

x	1	1	2	3	4
Allocation of random digits	0	1	2,3,4	5,6,7,8,	9

(b) A method similar to the above could be used in this case or alternatively the method indicated below can be adopted.

y	p_r	$p_r \times 1000*$ relative frequency	Cumulative relative frequency	Allocation of three digit random numbers
0	$\frac{7}{40} = 0.175$	175	175	001–175
1	$\frac{11}{40} = 0.275$	275	450	176–450
2	$\frac{9}{40} = 0.225$	225	675	451–675
3	$\frac{13}{40} = 0.325$	325	1000	676–999
Total	1	1000		together with 000

* The decision to multiply by 1000 was to ensure that we have integers in this column.

To sample from a Normal distribution

An alternative method of obtaining a random sample is to use a cumulative frequency polygon.
We first select a number at random along the CF scale to any degree of accuracy we choose. The corresponding value of X is then read off from the graph (Fig. 15.1).

This method can be used to select random samples from a continuous variate with a given pdf. It is also the basis for obtaining a random sample from $N(\mu, \sigma^2)$.

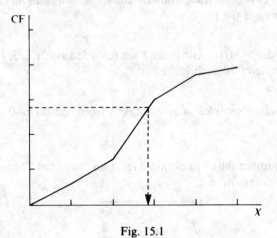

Fig. 15.1

Example 15.4

Show how to obtain a random sample from a Normal distribution with mean 5 and variance 9.
We first obtain a two digit random number from tables e.g. 74 and treat this as 0.74.
We then solve $\Phi(a) = 0.74$, using normal tables to obtain $z = 0.64$. Treating this as a standardised variable we solve

$$\frac{x - 5}{3} = 0.64,$$

to obtain

$$x = 6.92.$$

We can repeat this process as often as required.

Exercise 15.1

1. Use random number tables to obtain a random sample of 5 values from $N(3, 16)$.

2. Use random number table to draw a random sample of size 4 from
   (a) the distribution given below,

x	-2	1	0	1	2	3
p_r	$\frac{1}{15}$	$\frac{2}{15}$	$\frac{1}{3}$	$\frac{2}{15}$	$\frac{1}{15}$	$\frac{2}{3}$

   (b) a Binomial distribution with $n = 4$ $p = 1/3$.

3. Two players toss a die in turn. A player wins if his score is the same as that obtained by the previous player. A toss of a coin is used to decide who goes first. Simulate this game.

4. Choose a game of your own, e.g. roulette and show how random number tables can be used to simulate the game.

5. Values of a Poisson variate with mean 1 are recorded as 0, 1, 2, 3, 4 or more. Obtain a random sample of 10 such values.

6. Obtain 10 random samples of size 2 taken from the digits 0, 1, ..., 9 *without* replacement.

7. Use random number tables to obtain a random sample of 50 pairs of values from the frequency distribution

x	0	1	2	3
f	3	2	2	3

   assuming the samples are obtained (a) with replacement, (b) without replacement.
   In each case draw a histogram with your data and calculate the mean and variance. Compare these values with the theoretical values (cf. Section 12.4).

8. Describe how to use a random number table to simulate samples of the lengths of 10 men's left feet taken from a Normal population for which the mean is 0.27 m and the standard deviation 0.008 m.

Produce one such sample, giving the lengths to the nearest mm and using the following set of numbers which are assumed to be random.

46	24	07	02	33	98	09	13	40	68
57	11	47	50	00	29	15	29	83	21
37	30	42	29	38	97	22	70	52	92
94	84	38	82	84	18	84	90	39	09
40	05	36	93	72	18	19	88	98	62

AEB 1978

9. The methods that follow are suggested as ways of generating equiprobable random digits. In each case, find the frequency distribution of the digits generated and state, with your reasons, whether you consider that the method is satisfactory for its declared purpose.

(a) Throw two fair dice. Record the sum of the scores when the sum is 2–9; record the sum 10 as 0, and the sum 11 as 1; reject 12.

(b) Toss a fair penny four times. Record a head as 0 and a tail as 1 so that the result of a trial is four digits in order $abcd$, e.g. 0110. Interpret $abcd$ as the number $a \times 2^3 + b \times 2^2 + c \times 2 + d$ (e.g. 0110 → 6); record the number if it is 0–9, reject, it if it is greater than 9.

OXFORD

10. A clothing firm wishes to determine whether the average heights of adult Scots and Welsh males differ. It intends to send postcards (return postage paid) to each of 10 000 Scots males and 10 000 Welsh males drawn at random from names in telephone directories and to ask for these to be returned with the height of the recipient recorded. Criticise this method of sampling.

Describe briefly, giving reasons, how you would advise the firm to proceed, paying particular attention to the method of sampling and the subsequent analysis of the data.

MEI

15.3 Experiments

An important component of any course on Probability and Statistics is the experiment or study, particularly those involving the collection of information. Indeed in statistics we collect information about certain aspects of the world and try to form a model which 'fits' our experience. This model can then be used to make inferences about the course of future events.

Below there is a collection of experiments and exercises which can and should be used throughout the course. They can be modified to suit the needs of the reader as he progresses through the course and in some cases they can be extended to develop into projects if required. It is through these activities that the truth of the results given in this book becomes acceptable and perhaps more meaningful.

The order of the exercises follows the development of the subject but they can often be adapted to suit the needs and abilities of individuals where this is felt to be appropriate. In particular the data collected for one experiment should be retained so that as the reader develops his skills so more results can be obtained from the same data.

1. Select a sample of people from your school or college and ask them to tell you their favourite number between 1 and 9.

   If possible combine your results with others and form a frequency distribution. Draw a histogram of your results and superimpose the histogram you would expect to obtain if people had chosen randomly.

   Comment on your results.

   Can you think of ways in which your results might be unrepresentative of
   (a) the school or college?
   (b) the country?

2. Toss a coin until a head appears and record the total number of tosses required.

   Repeat this experiment at least 50 times. Form a frequency distribution, draw a histogram and calculate the mean and variance. If possible compare your results with others.

   You may be able to model this situation and hence compute the theoretical mean and variance.

3. Random-number tables are designed to represent values of a discrete variate which follows the uniform distribution for values between 0 and 9 inclusive. Examples of such numbers can be found on p. 304.

   The following are suggested methods for obtaining 'random' numbers.

   Toss a die and a coin together. If a 'head' is obtained record the score on the die minus two. (If a one is obtained on the die repeat the throw.) If a 'tail' is obtained then add four to the score, ignoring the result if a six is obtained.

   Show that this is bound to produce random numbers, assuming that both the die and coin are unbiased.

   Another method is to take a four digit number and using a calculator square the number and record the four middle digits. (If the square produces less than an eight digit display complete it by imagining a suitable number of zeroes at the beginning of the display.) Repeat the operation with this new four digit number. In this way a sequence of 'random' digits is obtained four at a time. These are called pseudo random numbers as they are not randomly generated. In fact one eventually regenerates the same sequence.

   Use this method to generate some random numbers, (at least 100) and draw a histogram to represent the data. Comment on your results.

   Another way to 'test' whether a sequence of numbers may be random is to count the number of digits between successive pairs of zeros (or ones, twos, etc.) The distribution should follow the geometric distribution and thus have a mean close to that obtained from theoretical considerations.

4. If a needle is dropped over a horizontal plane marked with parallel lines which are separated by a distance equal to the length of the needle there will be a definite probability that the needle crosses a line. This probability can be shown to be $2/\pi$. Test this by experiment and hence use your data to obtain an approximation to π.

5. Select two dominoes at random from a well-mixed set and establish whether they match or not. (Dominoes have two ends each of which displays a number of dots ranging from zero to six. Two dominoes match if they each have at least one end displaying the same number of dots.) Repeat the experiment as often as you can. (Combine your results with others if possible.)

   Estimate the probability that two dominoes chosen at random will match. Try to calculate the theoretical answer and compare with your own.

6. Obtain a large number of dried peas or other similar objects. Mark a known number, say 20, of the peas with ink. Extract the peas a few at a time and note the number n of marked peas and the number m of unmarked peas. Keep a running total of n and m.
   Now

$$\frac{n}{n+m} \simeq \frac{20}{N},$$

   where 20 peas are marked out of a total of N peas.
   Thus

$$N \simeq \frac{20(n+m)}{n}$$

   Compare your estimated value with the exact value of N.

   Explain how your method can be used to estimate the number of fish of a particular species in a lake. An alternative method is to select a sample of peas, say ten at a time and count the number r of marked peas, replacing the peas each time. The data can be approximated by the Binomial distribution, and so can be used to estimate the probability p, that a marked pea will be selected. This estimate can then be used to estimate the value of N.

7. Random-number tables can be used to simulate most games. For example, design a fruit machine and allocate digits to each possible outcome. Use random-number tables to determine the outcome of a 'trial'.

   If possible calculate the probability distribution of the variate which represents your 'winnings' and hence calculate the expected 'winnings' per game.

8. It has been stated that if X and Y are variates, then $E(X + Y) = E(X) + E(Y)$ regardless of whether X and Y are independent. However only if X and Y are independent can we be certain that $E(X \cdot Y) = E(X) \cdot E(Y)$.

   Toss two dice (or use random number tables) and let X represent the score on the first die and Y be the total score of the two dice. Clearly X and Y are not independent.

   Form a table of your results and use it to check the above facts.

9. To show that the Poisson distribution is the limiting form of the Binomial distribution when n is large and p is small. Take a sample of 10 pairs of consecutive random digits and record the number of pairs in the range 00–04 inclusive. By

grouping the results for consecutive samples obtain the number of pairs within the above range for samples of 20, 30 and 40 pairs.

Arrange each set of results in the form of a frequency distribution. Compute the corresponding expected frequency using the Poisson distribution with the same mean. Explain carefully why your method and results can be used to justify the above assertion.

10. On a plain square sheet of paper 10 cm × 10 cm ask a friend to place 25 points at random.

Superimpose a grid consisting of 25 2 × 2 cm squares and record the number of points in each square. Form a frequency distribution and compute the mean and variance. If these are approximately equal suggest a model for this distribution. (If they are not approximately equal what does this suggest about the randomness of the points?) Compare your model with the actual frequency distribution.
Explain
   (a) How you might use the method to test whether a certain species of plant grows at random in a particular field?
   (b) How you might test whether people can throw darts accurately with their left-hand? (Assuming they are right-handed).
   (c) How you might turn this experiment round and use it to test whether a set of so-called random digits are in fact random?

11. To show that the sum of two Poisson variates is in fact a Poisson variate you need to obtain a sequence of say 20 values of the variate with known mean and a further 20 values of another variate with known mean. This in turn will generate a further 20 values of the sum of these two variates. Compare your results in the form of a frequency distribution with those expected in theory.

12. To show that the sum of two Normally distributed variates is normal we can repeat the above experiments using a sequence of values from Normal variates.

Clearly this can be modified to show that the Central Limit Theorem is reasonable.

From Chapter 13 onwards the reader will find it increasingly easy to see how to use his new skills to make judgements about the truth or otherwise of certain statements. Great care should always be taken to ensure that the assumptions on which the results are based are reasonable. These are usually concerned with randomness and independence.

Chapter 16   The χ^2 distribution

16.1 Introduction

It has been noted previously that if $Z_1, Z_2, \ldots, Z_v$ are independent and each distributed $N(0, 1)$ then $Z = Z_1 + Z_2 + \ldots + Z_v$ is distributed $N(0, v)$.

We could not expect such an agreeable result for

$$\chi^2 = Z_1^2 + Z_2^2 + \ldots + Z_v^2$$

as in this case $\chi^2 \geqslant 0$ in all circumstances.

The χ^2 (chi squared, pronounced *ki squared*) distribution depends only on v, called the number of degrees of freedom. The notation $\chi^2(v)$ is used to indicate the number of degrees of freedom explicitly. Its probability density function is too complicated to consider here, but its mean is v, the number of degrees of freedom and its variance is $2v$.

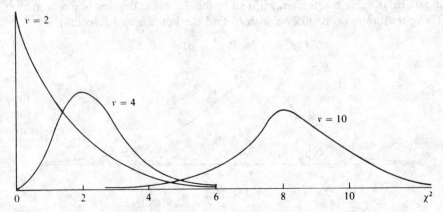

Fig. 16.1   The shape of the $\chi^2(v)$ distribution for $v = 2, 4$ and 10.

$\chi^2(1)$ and $\chi^2(2)$ are J-shaped, for other values the shape is similar to that of $\chi^2(10)$ though it tends to Normality as v increases (see Fig. 16.1). An alternative view of the definition of χ^2 is to consider a random sample $x_1, x_2, \ldots, x_n$ taken from $N(\mu, \sigma^2)$ then the statistic

$$\chi^2 = \frac{(x_1 - \mu)^2}{\sigma^2} + \frac{(x_2 - \mu)^2}{\sigma^2} + \ldots + \frac{(x_n - \mu)^2}{\sigma^2}$$

follows the $\chi^2 (n)$ distribution.

However, if μ is unknown and is estimated from the data we have

$$\chi^2 = \frac{(x_1 - \bar{x})^2}{\sigma^2} + \frac{(x_2 - \bar{x})^2}{\sigma^2} + \ldots + \frac{(x_n - \bar{x})^2}{\sigma^2}$$

$$= \frac{(n - 1)s^2}{\sigma^2},$$

where $s^2 = \Sigma(x - \bar{x})^2/n - 1$ is the unbiased estimator of σ^2 and since $E[(n - 1)s^2/\sigma^2]$ $= n - 1$ we are now considering a χ^2 $(n - 1)$ distribution. The loss of a 'degree of freedom' occurred because of the loss of independence of the variates. (Since $\bar{x}$ was calculated from the data a knowledge of $x_1, x_2, \ldots, x_{n-1}$ enables us to compute x_n and thus it is not independent!)

An immediate application of this result is to obtain confidence intervals for the variance of a Normal distribution. For suppose a random sample obtained from a Normal distribution is

$$0, \quad 3, \quad -1, \quad 1, \quad 0, \quad -1.$$

then

$$s^2 = 2.26$$

and so

$$\frac{5 \times 2.26}{\sigma^2} = \frac{11.3}{\sigma^2}$$

is a particular value taken from $\chi^2(5)$ since there are five degrees of freedom.

Using the tables on p. 305 we wish to find the limits α and β so that

$$P(\chi^2 \leqslant \alpha) = P(\chi^2 \geqslant \beta) = 0.025$$

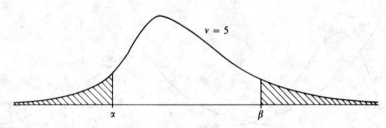

Fig. 16.2

Using the row corresponding to $v = 5$ these are

$$\alpha = 0.83 \quad \text{and} \quad \beta = 12.83.$$

Thus we are 95% confident that

$$0.83 \leqslant \frac{11.3}{\sigma^2} \leqslant 12.83$$

and so similarly confident that

$$\frac{11.3}{12.83} \leqslant \sigma^2 \leqslant \frac{11.3}{0.83}$$

$$\Rightarrow 0.88 \leqslant \sigma^2 \leqslant 13.61$$

Naturally we could improve this estimate by increasing the size of the sample.

16.2 Uses of the χ^2 distribution

Suppose two students were asked to toss a die about 100 times each and recorded their results as follows

	Score x	1	2	3	4	5	6
Student A	frequency	24	12	14	15	26	11
Student B	frequency	16	19	18	15	15	19

A natural question to ask is whether the data supports the hypothesis that the die is unbiased.

Intuitively we feel that if the differences between the *observed* frequencies and the *expected* frequencies (calculated on the basis of the hypothesis) were large then we would wish to reject the hypothesis.

It turns out that the statistic

$$\sum \frac{(O - E)^2}{E}$$

where the O's represent the observed frequencies and the E's represent the corresponding expected frequencies can be approximated by the χ^2 distribution. The approximation is generally considered acceptable if the expected frequencies are 5 or more.

For student A we have:

	1	2	3	4	5	6
O	24	12	14	15	26	11
E	17	17	17	17	17	17
$O - E$	7	-5	-3	-2	9	-6
$(O - E)^2$	49	25	9	4	81	36

The expected frequencies are calculated on the basis of the hypothesis and the total number of throws.

The expected frequencies are calculated on the basis of the hypothesis and the total number of throws.

Now
$$\sum \frac{(O - E)^2}{E} = \frac{1}{17}(49 + 25 + 9 + 4 + 81 + 36)$$

$$= \frac{204}{17}$$

$$= 12$$

As we shall see in the next section the number of degrees of freedom is 5.

On the basis of a 5% significance level we see from the table that the critical value, for a one-tailed test, is 11.07 and since our

$$\chi^2 = 12 > 11.07,$$

we can reject the hypothesis that the die is fair.

An inspection of the data suggests that the die may be biased in favour of the 1 and the 5.

On the other hand the corresponding statistic for student B's result is

$$\sum \frac{(O - E)^2}{E} = \frac{18}{17} = 1.06.$$

Again using the χ^2 (5) distribution we see that we would not reject the null hypothesis. Indeed a further glance at the table will indicate that the statistic is so small that an extremely unusual event has occurred, the data is too close to that expected! Since

$$\chi^2 = 1.06 < 1.15$$

where, assuming five degrees of freedom, only 5% of the distribution lies below 1.15. We are thus suspicious of the manner in which the data was obtained!

If after further investigation we were convinced that student B had obtained his results in a proper manner it is natural to ask whether we could improve on our tests by combining the results.

We note here that if X and Y are independent, χ^2 variates with degrees of freedom v_1, v_2 respectively then $X + Y$ is χ^2 $(v_1 + v_2)$.

The original definition of χ^2 makes this result reasonable.

Combining our results we find

$$\chi^2 = 12 + 1.06$$
$$= 13.06,$$

which is a value taken from χ^2 (10). For a 5% level the critical value is 18.31 and thus we do not reject the hypothesis that the die is fair on the basis of the evidence so presented.

A moment's reflection will show that the statistic $\Sigma(O - E)^2/E$ can be at most approximately χ^2 for, though χ^2 is a continuous distribution, the statistic $\Sigma(O - E)^2/E$ can only take a finite number of possible values for any chosen $N = \Sigma O$ and so is a discrete variate. We can however, justify the claim that it is approximately χ^2 in a particularly simple case.

Suppose we were interested in the number of sixes obtained when a die was tossed N times. On the null hypothesis that the die was unbiased we would have a Binomial situation. If we observed X sixes in the N trials then we have

	Frequency with which a 6 occurs	Frequency with which a 6 did not occur
Observed	X	$N - X$
Expected	$\dfrac{N}{6}$	$\dfrac{5N}{6}$

Then

$$\sum \frac{(O - E)^2}{E} = \frac{\left(X - \dfrac{N}{6}\right)^2}{\dfrac{N}{6}} + \frac{\left((N - X) - \dfrac{5N}{6}\right)^2}{\dfrac{5N}{6}}$$

$$= \frac{\frac{5}{6}\left(X - \frac{N}{6}\right)^2 + \frac{1}{6}\left(\frac{N}{6} - X\right)^2}{N \cdot \frac{1}{6} \cdot \frac{5}{6}}$$

$$= \frac{\left(X - \frac{N}{6}\right)^2}{N \cdot \frac{1}{6} \cdot \frac{5}{6}} \tag{*}$$

$$= \frac{(X - Np)^2}{Npq}, \quad \text{where } p = \frac{1}{6} \text{ and } q = 1 - p.$$

Provided N is large enough and Np, Nq are each greater than about 5, the statistic

$$\chi = \frac{X - Np}{\sqrt{Npq}}$$

is $N(0, 1)$, using the Normal approximation to the Binomial distribution and thus

$$\chi^2 = \frac{(X - Np)^2}{Npq} \quad \text{is } \chi^2(1)$$

and comparing (*) with χ^2 we see that in this case $\Sigma(O - E)^2/E$ is approximately $\chi^2(1)$.

The number of degrees of freedom

In the statistic $\Sigma(O - E)^2/E$ the E's are calculated on the basis of the null hypothesis together, possibly, with other constraints. In the example above both student A's and student B's expected frequencies E were calculated using the constraint that $\Sigma E = \Sigma O = 102$.

This constraint loses one degree of freedom (if the first $n - 1$ E's are chosen independently then the last E is fixed by this constraint).

In general we have

> *Number of degrees of freedom* = (number of cells in the table) − (the number of constraints on the expected frequencies)

Example 16.1

The number of cosmic particles recorded in 60 successive minutes was found to be

Number of particles x	0	1	2	3	4	5	6 or more
Frequency f	20	18	13	6	2	1	0

Is there evidence to support the claim that the emission of cosmic particles occurs randomly?

In this problem the null hypothesis is that the data came from a Poisson distribution. In order to obtain the expected frequencies we need to estimate the mean from the data.

$\bar{X} = 1.25$ and thus, using $P(1.25)$, the associated probabilities and the corresponding expected values with $N = 60$ are

	0	1	2	3	4	5	6 or more
Probabilities	0.2865	0.3581	0.2238	0.0933	0.0291	0.0073	0.0019
Expected frequencies	17.2	21.5	13.4	5.6	1.7	0.4	0.1

As some of the expected frequencies are smaller than 5 we group the last four cells together. (If there is a choice in the manner of grouping it is better to arrange to keep the expected values as nearly equal as possible). Our amended table is then

Number of particles	0	1	2	3 or more
O frequency	20	18	13	9
E frequency	17.2	21.5	13.4	7.8
$O - E$	2.8	-3.5	-0.4	1.2
$(O - E)^2$	7.84	12.25	0.16	1.44
$\dfrac{(O - E)^2}{E}$	0.4558	0.5698	0.0119	0.1846

$$\sum \frac{(O - E)^2}{E} = 0.4558 + 0.5698 + 0.0119 + 0.1846$$
$$= 1.221$$

In this example the value 1.2221 is taken from χ^2 (2). We have four cells, and have two degrees of freedom. One is lost because of the constraint $\Sigma E = 60$ and the other because we must make $\bar{E} = 1.25$.

For a 5% level of significance we see from tables that the critical value is 5.99 and as

$$\chi^2 = 1.2221 < 5.99,$$

we cannot reject the null hypothesis. Thus the evidence supports the view that the data occurred at random.

Exercise 16.1

(test at the 5% level of significance)

1. Establish a 99% confidence interval for the variance of a Normal distribution using the random sample below

16.4	22.0	11.8	15.0	19.6	18.2

2. A die was tossed 60 times with the following results.

Score	1	2	3	4	5	6
Frequency	11	13	9	14	6	7

Test whether the evidence supports the hypothesis that the die is unbiased.

3. Some digits were written down apparently at random and then grouped into the following frequency table.

Digits	0	1	2	3	4	5	6	7	8	9
Frequency	16	26	17	24	15	27	13	24	18	20

Test whether the numbers could indeed be random.

4. In a sample of the electorate 240 said they would vote Conservative, 280 Labour and 480 would vote for neither of these parties. In the actual election 28% of the electorate voted Conservative and 30% voted Labour. Do you think that the sample was representative of the voting habits of the electorate?

5. The number of thunderstorms over a particular location was recorded each month for a period of 5 years.

Number of thunderstorms per month	0	1	2	3	4 or more
Frequency	33	23	3	1	0

Test whether the data conforms to the Poisson distribution.

6. Three coins were tossed 60 times and the number of heads recorded, with the following results.

Number of heads	0	1	2	3
Frequency	4	29	18	9

Do you think the coins were unbiased?

7. A sample of values of the continuous variate X were chosen randomly and classified as follows.

x	0–1	1–2	2–3
f	14	33	13

Is there evidence to suggest that X has a pdf of the form
(a) a rectangular distribution over 0 to 3,

(b) $\begin{cases} kx\,(3-x) & 0 \leqslant x \leqslant 3 \\ 0 & \text{otherwise} \end{cases}$.

8. Prove that $\Sigma[(O - E)^2/E] = \Sigma[O^2/E] - N$, where $N = \Sigma O = \Sigma E$.

9. Six coins were tossed 640 times and the number of heads recorded.

Number of heads	0	1	2	3	4	5	6
Frequency	18	66	167	185	142	55	7

(a) Test the hypothesis that the coins are unbiased.
(b) Test the hypothesis that the Binomial distribution is a suitable model. (Hint: assume a constant probability p of landing heads for each of the coins. Then use the data to estimate p).

10. Suppose the probability that a coin lands heads is thought to be p and yet when tossed N times r heads are obtained. Show that the statistic can be approximated by the χ^2 distribution with one degree of freedom.

11. A sample of two children families was investigated with the following results

	Two boys	One boy and one girl	Two girls
Number of families	61	93	46

Is the data consistent with the hypothesis that boys and girls are equally likely?

12. The number of empty lemonade bottles returned to a store during a week was recorded as follows

Monday	Shop closed
Tuesday	84
Wednesday	48
Thursday	93
Friday	87
Saturday	90
Sunday	Shop closed

(a) Test the hypothesis that shoppers are equally likely to return bottles on any of the days for which the shop is open.

(b) Test the assertion that only half as many bottles will be returned on a Wednesday as on any of the other days for which the shop is open.

16.3 Contingency tables

When data is classified in two different ways the resulting table is called a *contingency table*.

The table Fig. 16.3 represents the grades obtained by 100 students in both Mathematics and French.

		French grades			
		A	B	C	Totals
Maths grades	A	9	14	10	33
	B	18	17	12	47
	C	6	11	3	20
	Totals	33	42	25	100 = sum of totals

Fig. 16.3 The table has been extended by calculating the (marginal) totals.

So, for example, 12 students gained a B grade in Mathematics and a C grade in French. Such a table is called a 3 × 3 contingency table (3 rows by 3 columns) though they come in all sizes. An $m \times n$ table would consist of m rows and n columns.

A natural question to ask is whether there is evidence that the Mathematics grades are independent of the French grades.

On the basis of the null hypothesis that the grades are independent we can calculate the expected frequencies and thus compute the value of χ^2 as follows:

From the table we see that the probability of obtaining a grade B in Mathematics is

47/100. Similarly the probability of obtaining a C in French is $25/100 = 1/4$. Thus if we assume that these events are independent the probability of obtaining a C in Mathematics and a B in French is

$$\frac{47}{100} \times \frac{25}{100}$$

So the expected frequency in the (B, C) cell would be

$$100 \times \frac{47}{100} \times \frac{25}{100} = \frac{47 \times 25}{100} = 11.75$$

In general the expected number in the cell (α, β) is

$$\frac{\text{product of margined totals associated with cell}}{\text{sum of totals}}$$

We thus obtain the expected frequencies as

		French		
		A	B	C
	A	$\frac{33 \times 33}{100}$	$\frac{42 \times 33}{100}$	$\frac{25 \times 33}{100}$
Maths	B	$\frac{33 \times 47}{100}$	$\frac{42 \times 47}{100}$	$\frac{25 \times 47}{100}$
	C	$\frac{33 \times 20}{100}$	$\frac{42 \times 20}{100}$	$\frac{25 \times 20}{100}$

which reduces to

		French		
		A	B	C
	A	10.89	13.86	8.25
Maths	B	15.51	19.74	11.75
	C	6.60	8.40	5.00

We now compute χ^2

$$\chi^2 = \sum \frac{(O - E)^2}{E}$$

$$= \frac{(9 - 10.89)^2}{10.89} + \frac{(14 - 13.86)^2}{13.86} + \ldots$$

$$\ldots + \frac{(3 - 5.00)^2}{5.00}$$

$$= 3.145.$$

We now need to establish the number of degrees of freedom. An inspection of the expected frequencies will show that the marginal totals are the same as the original ones; and so there is more than one constraint on the evaluation of these expected frequencies. Indeed if we imagine choosing the four expected frequencies for the cells shaded in the diagram below, then the other five cells can be computed using the known marginal totals. We thus have four degrees of freedom.

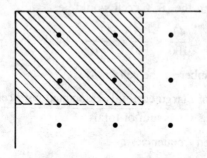

Similarly an $(m \times n)$ contingency table will have $(m - 1) \times (n - 1)$ degrees of freedom.

Comparing our $\chi^2 = 3.145$ with the critical value at a 5% level of significance using 4 degrees of freedom we see that we cannot reject the hypothesis that the grades are independent.

Yates' correction

In the special case of a 2×2 contingency the number of degrees of freedom is reduced to one (and thus the square root of χ^2 will be a normal variate!). In these circumstances the approximation to the continuous χ^2 distribution is improved if Yates' correction is applied. (This is similar to the correction applied when using the Normal distribution as an approximation to the Binomial distribution.) The corrected formula is

$$\sum \frac{(|O - E| - \frac{1}{2})^2}{E}$$

Each difference is reduced in absolute value by $1/2$, thus if $O - E = 3.6$ we would use 3.1 and if $O - E = -3.6$ we would use $+3.1$.

It is particularly important in 2×2 tables that the expected frequencies should be 5 or more.

Example 16.2

During a trial for a new cold treatment 70 volunteers out of a total of 200 were treated at the beginning of the trial. During this period the number contracting a cold were as follows.

	Cold	No cold
Treated	10	60
Not treated	26	104

Test the hypothesis that the cold treatment is ineffective.

On the basis of the hypothesis we can use independence in the usual way to calculate the expected frequencies.

<table>
<tr><td colspan="4">Observed frequencies</td><td colspan="3">Expected frequencies</td></tr>
<tr><td></td><td>Cold</td><td>No cold</td><td></td><td></td><td>Cold</td><td>No cold</td></tr>
<tr><td>Treated</td><td>10</td><td>60</td><td>70</td><td>Treated</td><td>12.6</td><td>57.4</td></tr>
<tr><td>not</td><td></td><td></td><td></td><td>not</td><td></td><td></td></tr>
<tr><td>treated</td><td>26</td><td>104</td><td>130</td><td>treated</td><td>23.4</td><td>106.6</td></tr>
<tr><td></td><td>36</td><td>164</td><td>200</td><td></td><td></td><td></td></tr>
</table>

(We apply Yates correction since we are dealing with a 2 × 2 contingency table.)

$$\chi^2 = \sum \frac{(|O - E| - \frac{1}{2})^2}{E}.$$

$$= \frac{(-2.1)^2}{12.6} + \frac{(2.1)^2}{57.4} + \frac{(2.1)^2}{23.4} + \frac{(-2.1)^2}{106.6}$$

$$= 0.657.$$

Since the critical value in $\chi^2(1)$ at the 5% level of significance is 3.84 there is no reason to reject the hypothesis that the cold treatment is ineffective. Applying Yates' correction makes it harder to reject the null hypothesis!

Exercise 16.2

1. Test the hypothesis that possessing a colour television set is independent of the type of district in which a family lives, using the data below.

Location	Owner	Non-owner
City	58	25
Town	26	13
Country	17	21

2. An experiment was repeated three times with the following results

	1st experiment	2nd experiment	3rd experiment
$\chi^2(3)$	7.1	6.9	6.7

Has a significant result been obtained?

3. Some students decided to test the assertion that one cannot tell the difference between butter and a particular brand of margarine. They recorded their results as follows.

	Could tell	Could not tell
Male	16	14
Female	21	9

Do you think that there is any significant difference between the ability of males and females in distinguishing between butter and margarine?

4. The performance of two machines were investigated with the following result.

		Number of defectives	Number of articles produced
Machine	A	19	401
	B	31	599

Test the hypothesis that there is no significant difference between the performance of these two machines.

5. Use the following data to test whether there is any significant difference in voting pattern between ages.

		Vote Conservative	Vote Labour
	21–29	31	39
Age	30–49	86	114
	50–65	65	35

6. In order to test the claim that red-haired women tend to be more bad tempered than others the following data was collected.

		Colour of hair	
		Red	Other than red
Temper	Good	6	42
	Bad	16	84

What conclusions can you draw?

Chapter 17 Correlation

17.1 Introduction

In this chapter we shall consider the interdependence of two random variables X and Y. We suppose that data is collected in pairs, one a value of X, the other a corresponding value of Y, taken from the same individual or at the same time. For example the weight X and height Y of a sample of individuals was found to be

Weight kg	67.3	68.3	70.9	70.0	65.9	68.2	71.8	74.0	71.9	69.5	X
Height m	1.74	1.69	1.82	1.79	1.61	1.83	1.95	1.92	1.81	1.75	Y

Data of this type occurs quite naturally, particularly in Biology and the Social Sciences. We might, for example, be interested in the temperature and height above sea level at various locations, the IQ and reading ages of various students, or the speed of an object at various times.

One problem, to be considered in the next chapter, is whether we can reasonably predict the value of one of the variables from a knowledge of the other. Another question, which we consider in this chapter is whether there is a relationship between the variables at all. For example, does the data above provide evidence that tall people tend to be heavy and short people less so? If there is a linear relationship between the variables then we shall say that they are *linearly correlated* (or simply *correlated*.)

The natural first step is to plot the data on a graph (called a *scatter diagram*) (see Fig. 17.1).

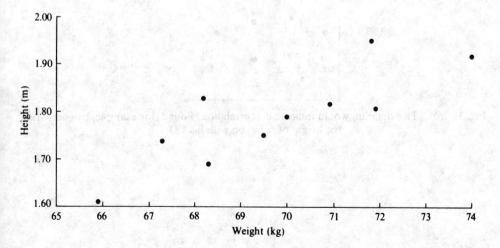

Fig. 17.1 A *scatter diagram* to show the height and weight of a sample of people.

An inspection of the data displayed in this form suggests that there may indeed be a correlation between the two variables. In this case a linear relationship with an upward slope, which is referred to as positive correlation. If the underlying linear relationship has a downward slope, then the variables are said to be negatively correlated. If there is no evident linear relationship, perhaps because the variables are independent or are related in a non-linear way, then we shall say the variables are uncorrelated, meaning not linearly correlated.

The reason the data diverges from the underlying relationship is due to the fact that one or both of the variables has a probability distribution due to errors of measurement, chance fluctuations or the effect of other factors.

Other typical scatter diagrams are shown in Fig. 17.2(a)–(d).

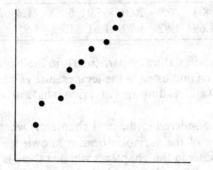

Fig. 17.2(a)   This is an example of near perfect correlation, obtained for example when the speed of a particle moving under constant accelaration is measured at various times. Errors are introduced whenever a measurement takes place, so the data cannot be expected to lie precisely on a straight line.

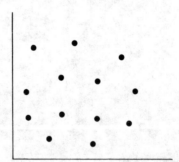

Fig. 17.2(b)   This diagram would indicate no correlation. Found, for example, by comparing the height of a person with his I.Q.

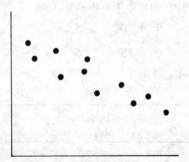

Fig. 17.2(c)   This diagram indicates a negative correlation. Found, for example, by comparing the temperature and the height above sea-level at various locations.

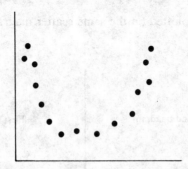

Fig. 17.2(d)   This is an example of a non-linear relationship between the variables. The methods we develop in this chapter presume a linear relationship, and care must be taken to avoid applying the results in those cases involving a non-linear relationship.

17.2 The product moment correlation coefficient

As in previous chapters we have not been content simply with the display of data but have sought a numerical measure of the concept of interest.

In this case we would like a number which will distinguish between the various degrees of correlation from 'positive' through 'no correlation' to 'negative'.

As an aid to developing such a measure we consider the Maths and Physics marks of a group of students

| Maths | 12 | 10 | 8 | 12 | 6 | 10 | 12 | 8 | 12 | 10 | X |
| Physics | 21 | 15 | 12 | 15 | 12 | 18 | 12 | 15 | 18 | 12 | Y |

A scatter diagram would indicate some positive correlation, which is perhaps more than a cursory inspection of the data would suggest. In part the difficulty is in identifying 'high' (above the mean) and 'low' (below the mean) marks. This difficulty can be resolved by standardising the marks. That is, we find the number of standard deviations each mark is from its mean.

It is easy to check that in this (rather artificial) case

$$\bar{X} = 10 \qquad S_x = 2$$

$$\bar{Y} = 15 \qquad S_y = 3$$

giving the standardised marks as

Maths	1	0	−1	1	−2	0	1	−1	1	0	z_x
Physics	2	0	−1	0	−1	1	−1	0	1	−1	z_y

where $\quad Z_X = \dfrac{X - \bar{X}}{S_X} \quad$ and $\quad Z_Y = \dfrac{Y - \tilde{Y}}{S_Y}$

These two sets of data are plotted on the same scatter diagram (Fig. 17.3).

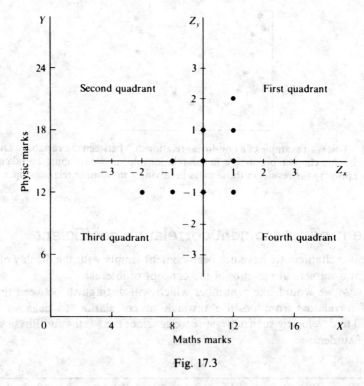

Fig. 17.3

The effect of standardising the marks has been to draw new axes through the points $(\bar{X}, \bar{Y})$ with new scales (units of standard deviations) rather than use the essentially arbitrary scale of marks. Such a construction does highlight an important property, namely that the points in the first and third quadrants are the ones which contribute to a positive correlation and those in the other two quadrants to a negative correlation.

A numerical measure of the degree of correlation could be taken to be

$$\Sigma z_x z_y$$

If a particular point (z_x, z_y) is in the first or third quadrant, then z_x and z_y are both positive or both negative and in either case the product is positive. If the point is in the second or fourth quadrant then the product will be negative. Thus if most points were in the first and third quadrants $\Sigma z_x z_y$ would tend to be positive, and if most of the points are in the second and fourth quadrants $\Sigma z_x z_y$ would tend to be negative. In the case of no correlation the points would be more evenly distributed between the quadrants and $\Sigma z_x z_y$ would be close to zero.

One major fault with this measure is that the magnitude of $\Sigma z_x z_y$ depends to some extent on the number of items of data, so it would be difficult to use this measure in comparison with that obtained from a larger group of students. A simple solution is to average the products; and so we adopt as our measure of correlation the *product moment correlation coefficient r* given by

$$r = \frac{1}{n}\Sigma z_x z_y,$$

where n represents the number of pairs of data and z_x, z_y the standardised scores.

17.3 Alternative forms of the product moment correlation coefficient

We seek an alternative form of the above equation suitable for use in calculations. Now

$$r = \frac{1}{n}\Sigma z_x z_y$$

$$= \frac{1}{n}\Sigma\left(\frac{x-\bar{x}}{S_x}\right)\left(\frac{y-\bar{y}}{S_y}\right)$$

$$= \frac{\frac{1}{n}\Sigma(x-\bar{x})(y-\bar{y})}{S_x S_y}$$

$\frac{1}{n}\Sigma(x-\bar{x})(y-\bar{y})$ is called the *Covariance of the sample* and can be written as

$$S_{xy} = \frac{1}{n}(\Sigma xy - n\bar{x}\bar{y})$$

So

$$r = \frac{S_{xy}}{S_x \cdot S_y}$$

(See Exercise 17.1 question 4 for a proof. Compare also with Section 10.4.)
Similarly S_x the standard deviation of the sample of X values can be written as

$$\sqrt{\frac{1}{n}(\Sigma x^2 - n\bar{x}^2)}$$

Thus

$$r = \frac{\frac{1}{n}(\Sigma xy - n\bar{x}\bar{y})}{\sqrt{\frac{1}{n}(\Sigma x^2 - n\bar{x}^2)\cdot\frac{1}{n}(\Sigma y^2 - n\bar{y}^2)}} = \boxed{\frac{\overline{xy} - \bar{x}\bar{y}}{\sqrt{(\overline{x^2} - \bar{x}^2)(\overline{y^2} - \bar{y}^2)}}}$$

Cancelling $\dfrac{1}{n}$ and writing $\bar{x} = \dfrac{\Sigma x}{n}$ and $\bar{y} = \dfrac{\Sigma y}{n}$ we obtain

Definition 17.1

> The product moment correlation coefficient is
>
> $$r = \frac{\Sigma xy - \dfrac{\Sigma x\,\Sigma y}{n}}{\sqrt{\left(\Sigma x^2 - \dfrac{(\Sigma x)^2}{n}\right)\left(\Sigma y^2 - \dfrac{(\Sigma y)^2}{n}\right)}}$$

which is the most useful form for use in calculations.

Coding
Before we use this formula to calculate r for the sample of heights and weights given above we note a useful fact. If the values of X and Y are coded by say $u = ax + b$ and $v = a'y + b'$, this will have no effect on the value of r. See exercise 17.1 for further details, but this is simply a restatement of the fact that, since r is measured in standardised variables, the scale of X and Y can be altered without affecting r.

X	Y	$U = X - 70$	$V = 100Y - 180$	U^2	V^2	UV
67.3	1.74	-2.7	-6	7.29	36	16.2
68.3	1.69	-1.7	-11	2.89	121	18.7
70.9	1.82	0.9	2	0.81	4	1.8
70.0	1.79	0	-1	0	1	0
65.9	1.61	-4.1	-19	16.81	361	77.9
68.2	1.83	-1.8	3	3.24	9	-5.4
71.8	1.95	1.8	15	3.24	225	27.0
74.0	1.92	4	12	16	144	48.0
71.9	1.81	1.9	1	3.61	1	1.9
69.5	1.75	-0.5	-5	0.25	25	2.5
		$\Sigma u = -2.2$	$\Sigma v = -9$	$\Sigma u^2 = 54.14$	$\Sigma v^2 = 927$	$\Sigma uv = 188.6$

Substituting into

$$r = \frac{\Sigma uv - \dfrac{\Sigma u \, \Sigma v}{n}}{\sqrt{\left(\Sigma u^2 - \dfrac{(\Sigma u)^2}{n}\right)\left(\Sigma v^2 - \dfrac{(\Sigma v)^2}{n}\right)}}, \quad \text{where} \begin{cases} n = 10 \\ \Sigma u = -2.2 \\ \Sigma v = -9 \\ \Sigma u^2 = 54.14 \\ \Sigma v^2 = 927 \\ \Sigma uv = 188.6 \end{cases}$$

$$= \frac{188.6 - 1.98}{\sqrt{(54.14 - 0.484)(927 - 8.1)}}$$

$$= 0.84$$

note again that coding has had no effect, on the value of r.

17.4 Interpretation of the correlation coefficient

In Section 10.4 the product moment correlation coefficient was defined for two random variables as

$$\rho = \frac{\text{Cov}(X, Y)}{\sigma_x \sigma_Y}$$

and r can be used as an (unbiased and consistent) estimator of ρ provided that a random sample is chosen. Strictly, we should use the formula

$$r = \frac{s_{xy}}{s_x s_y},$$

where

$$s_x = \sqrt{\frac{\Sigma(x - \bar{x})^2}{n - 1}} \quad \text{and} \quad s_y = \sqrt{\frac{\Sigma(y - \bar{y})^2}{n - 1}}$$

are the unbiased estimates of the population standard deviation, and $s_{xy} = \Sigma(x - \bar{x})(y - \bar{y})/n - 1$ is an estimate of Cov (X, Y), the population covariance, but the factor $1/(n - 1)$ makes no difference to the value of r since it cancels. We should, however use the above formula when we wish to estimate the covariance of the population from a small random sample.

A very useful property of r (and ρ) is that no matter what sample we consider

$$-1 \leqslant r \leqslant 1 \qquad \text{(see Exercise 17.1)}$$

and r is a very suitable measure of correlation since large absolute values of r indicate a high degree of correlation and values close to zero represent no linear correlation.

A large absolute value of r, indicating correlation, does not imply causality. For example, we found a degree of positive correlation ($r = 0.84$) between the height and weight of a sample of people, but this does not mean that being heavy causes one to grow tall. It would be reasonable to assume a high positive correlation between the number of teachers and the number of unsolved crimes over a period of time. But this would simply be reflecting an increase in both during that time period. It should also be noted that r measures only the strength of a linear relationship between the variables. If data were distributed as in Fig. 17.2(d) then r would be close to zero, implying little linear correlation, rather than the variables being independent. It is for this reason that drawing a scatter diagram is important.

Exercise 17.1

1. Construct a scatter diagram to illustrate the following set of bivariate data, stating how the two variables are correlated.

X	0	2	4	6	8
Y	40	30	20	10	0

2. Calculate the value of r for the following sets of data.

(a) X	1	2	3	4	5
Y	3	5	7	9	11

(b) X	1	1	−1	−1
Y	1	−1	1	−1

(c) X	−3	−2	−1	0	1	2	3
Y	5	0	−3	−4	−3	0	5

In each case draw a scatter diagram and comment on your results.

3. Select two convenient measures applicable to a sample of people such as their arm length and their shoe size. First estimate and then calculate r for your data.

4. Show that $\Sigma(x - \bar{x})(y - \bar{y}) = \Sigma xy - n\bar{x}\,\bar{y}$, where the summation is taken over the bivariate data.

X	x_1	x_2	$x_3 \ldots x_n$
Y	y_1	y_2	$y_3 \ldots y_n$

5. If $U = aX + b$ and $V = a'Y + b'$ show that

$$\text{Cov}(U, V) = aa'\,\text{Cov}(X, Y)$$

and hence

$$\rho = \frac{\text{Cov}(U, V)}{\sigma_u \sigma_v} = \frac{\text{Cov}(X, Y)}{\sigma_x \sigma_y}.$$

6. By expanding $\Sigma[(x - \bar{x}) + \lambda(y - \bar{y})]^2$ as a quadratic function of λ and noting that the expression is non-negative for all values of λ deduce that

$$S_{xy}^2 \leqslant S_x^2 S_y^2$$

and hence

$$-1 \leqslant r \leqslant 1$$

7. Define the product moment correlation coefficient r. What is implied by values of r close to $+1$ or close to -1? If the two variables are distributed independently, what does this imply about the value of r?

The following figures relate to the tensile strengths of 12 samples of steel at two different temperatures, T_1 and T_2. Calculate the product moment correlation coefficient between the tensile strengths and comment on your result.

MEI

Sample number	1	2	3	4	5	6	7	8	9	10	11	12
Tensile strength												
At T_1	55	50	65	55	60	60	70	35	60	50	70	65
At T_2	35	25	40	35	45	40	45	15	45	35	45	40

8. State the effect on the product moment correlation coefficient between two variables x and y of (i) changing the origin for x and (ii) changing the units of x. The following table gives the daily output of the substance creatinine from the body of each of ten nutrition students together with the student's body mass.

Output of creatinine (grammes)	1.32	1.54	1.45	1.06	2.13	1.00	0.90	2.00	2.70	0.75
Body mass (kilograms)	55	48	55	53	74	44	49	68	78	51

Draw a scatter diagram for the data.

Calculate, correct to two decimal places, the product moment correlation coefficient.

Comment on any relationship which is indicated by the scatter diagram and the correlation coefficient.

JMB

9. Find the value of the product moment correlation coefficient for a set of bivariate data if

(a) $S_{xy} = 29$, $S_x = 15$, $S_y = 24$;

(b) $S_{xy} = -5$, $S_x^2 = 14$, $S_y^2 = 12$;

(c) $\Sigma xy = 2.2$, $\Sigma x = 12$, $\Sigma y = 10$, $\Sigma x^2 = 180$, $\Sigma y^2 = 130$, and $n = 10$.

17.5 Spearman's rank correlation coefficient

Suppose some students gained the following grades at A level in mathematics and physics.

Maths	A	O	B	F	E	C	X
Physics	C	B	A	O	O	B	Y

We would like to be able to establish whether there is any degree of correlation between these two sets of grades. In this case, however, we do not have a numerical scale. What we can do is *rank* the grades in order of merit. In other words, we order the students grades in each subject from best to worst. (In other situations we might mean highest to lowest or largest to smallest etc.)

Thus in the case of mathematics the ranks are

$$\textit{Mathematics} \quad 1 \quad 5 \quad 2 \quad 6 \quad 4 \quad 3 \quad X$$

In the case of the physics marks a problem arises since there are *tied ranks*. The second and sixth student each have the same grade as do the fourth and fifth students. In this situation the ranks allotted are the mean of the ranks available and so we have

$$\textit{Physics} \quad 4 \quad 2\tfrac{1}{2} \quad 1 \quad 5\tfrac{1}{2} \quad 5\tfrac{1}{2} \quad 2\tfrac{1}{2} \quad Y$$

We now use the ranked data to compute the correlation coefficient using the formula for r. When substituting ranks rather than numerical data the resulting coefficient is called *Spearman's rank correlation coefficient* and will be denoted by r_S.

It should be noted that even when the original data is numerical and thus r can be computed it is possible to rank the data and compute r_S sometimes resulting in a considerable saving of time.

Example 17.1
Calculate Spearman's rank correlation coefficient for the above data.

X	Y	$U = X - 3$	$V = Y - 3$	U^2	V^2	UV
1	4	-2	1	4	1	-2
5	2.5	2	-0.5	4	0.25	-1
2	1	-1	-2	1	4	2
6	5.5	3	2.5	9	6.25	7.5
4	5.5	1	2.5	1	6.25	2.5
3	2.5	0	-0.5	0	0.25	0
		$\Sigma u = 3$	$\Sigma v = 3$	$\Sigma u^2 = 19$	$\Sigma v^2 = 18$	$\Sigma uv = 9$

$$r_S = \frac{\Sigma uv - \dfrac{\Sigma u \, \Sigma v}{n}}{\sqrt{\left(\Sigma u^2 - \dfrac{(\Sigma u)^2}{n}\right)\left(\Sigma v^2 - \dfrac{(\Sigma v)^2}{n}\right)}}$$

$$= \frac{9 - 1.5}{\sqrt{(19 - 1.5)(18 - 1.5)}}$$

$$= 0.44$$

Simplified formula for r_S

In the case that there are no tied ranks we can simplify the formula for r_S to

$$r_S = 1 - \frac{6\Sigma(x - y)^2}{n(n^2 - 1)}$$

$$= 1 - \frac{6\Sigma d^2}{n(n^2 - 1)},$$

where $d = x - y$ the rank differences, which as we shall see is a very simple formula to use.

In the general case of n pairs of data both X and Y will take the values 1 to n in some order.

Thus

$$\Sigma x = \Sigma y = \frac{n}{2}(n + 1), \qquad \text{the sum of an arithmetic series,}$$

and

$$\Sigma x^2 = \Sigma y^2 = \frac{n}{6}(n + 1)(2n + 1) \qquad \text{the sum of the first } n \text{ square numbers. It is this result which demands no tied ranks.}$$

So

$$\Sigma x^2 - \frac{(\Sigma x)^2}{n} = \Sigma y^2 - \frac{(\Sigma y)^2}{n} = \frac{n}{6}(n + 1)(2n + 1) - \frac{n^2(n + 1)^2}{4n}$$

$$= \frac{n(n^2 - 1)}{12}.$$

Further, since

$$ab = \frac{1}{2}(a^2 + b^2 - (a - b)^2)$$

$$\Sigma xy = \frac{1}{2}(\Sigma x^2 + \Sigma y^2 - \Sigma(x - y)^2)$$

$$= \frac{n}{6}(n + 1)(2n + 1) - \frac{1}{2}\Sigma d^2$$

So

$$\Sigma xy - \frac{\Sigma x \, \Sigma y}{n} = \frac{n(n^2 - 1)}{12} - \frac{1}{2}\Sigma d^2.$$

Substituting into the formula we obtain

$$. r_S = \frac{\dfrac{n(n^2 - 1)}{12} - \dfrac{1}{2}\Sigma d^2}{\dfrac{n(n^2 - 1)}{12}}$$

$$= 1 - \frac{6\Sigma d^2}{n(n^2 - 1)}.$$

Example 17.2
Calculate Spearman's rank correlate coefficient for the data given at the beginning of this chapter, where r was found to be 0.84.

X	Y	X ranks	Y ranks	$X - Y = d$	d^2
67.3	1.74	9	8	1	1
68.3	1.69	7	9	-2	4
70.9	1.82	4	4	0	0
70.0	1.79	5	6	-1	1
65.9	1.61	10	10	0	0
68.2	1.83	8	3	5	25
71.8	1.95	3	1	2	4
74.0	1.92	1	2	-1	1
71.9	1.81	2	5	-3	9
69.5	1.75	6	7	-1	1
					$\Sigma d^2 = 46$

$$r_S = 1 - \frac{6 \times 46}{10 \times 99}$$

$$= 0.72$$

compare this with the calculation required for r on p. 254.

Significance of the product moment correlation coefficient

We have already stated that a value of r close to 1 or -1 indicates a high degree of correlation and if close to zero little or no (linear) correlation. Suppose, however, we thought $\rho = 0$, that is, there is no correlation between the random variables; then we would expect r to be small if calculated from a sample chosen at random. If we did obtain a large absolute value of r we would have to abandon our hypothesis that $\rho = 0$.

The critical values depend on

(a) the size of the sample,
(b) the alternative hypothesis (is it a one-tailed or two-tailed test?),
(c) the distribution of X and Y.

If the random variables are assumed to be jointly Normally distributed then the table in Fig. 17.4 can be used to test

$$\text{NH} : \rho = 0$$

$$\text{against} \quad \text{AH} : \rho \neq 0 \quad (\Rightarrow \text{two-tailed})$$

$$\text{or} \quad : \rho > 0 \quad (\rho < 0) \quad (\Rightarrow \text{one-tailed}).$$

It should be noted that to say X and Y are jointly Normally distributed is stronger than claiming that both X and Y are Normal. Since, in addition, we require that for each value x of X the values of Y which can associate with this x are themselves Normally distributed. The same may be said for Y.

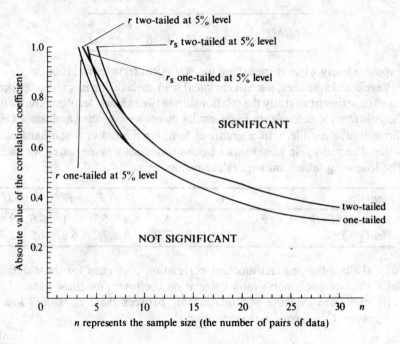

Fig. 17.4

The graph shows the approximate values of r and r_S required to give a significant result for various sample sizes for both a one-tailed and a two-tailed test at the 5% level.

Note that for samples of size 10 or more there is little difference in the critical values of each of the coefficients. For example, if a correlation coefficient of 0.5 is obtained from a sample of size 15, then, assuming joint normal variables, the NH : $\rho = 0$ may be rejected in favour of the AH : $\rho > 0$ (but not in favour of the AH : $\rho \neq 0$).

Exercise 17.2

1. Eight different makes of valves are tested and ranked according to efficiency. The results, together with their prices are shown below. Do you think that price is a good guide to performance?

Price (£)	13	10	12	7	16	11	5	14
Rank	2	7	3	6	4	5	8	1

2. A sample of student's essays were marked by two teachers independently. The resulting ranks are shown below. Calculate the rank correlation coefficient and state any conclusion you draw from it.

Teacher								
A	5	8	1	6	2	7	3	4
B	7	4	3	1	6	8	5	2

3. Explain clearly what is meant by the statistical term 'correlation'.

Vegboost Industries, a small chemical firm specializing in garden fertilizers, set up an experiment to study the relationship between a new fertilizer compound and the yield from tomato plants. Eight similar plants were selected and treated regularly throughout their life with x grams of fertilizer diluted in a standard volume of water. The yield y, in kilograms, of good tomatoes was measured for each plant. The following table summarizes the results.

Plant	A	B	C	D	E	F	G	H
Amount of fertilizer x (g)	1.2	1.8	3.1	4.9	5.7	7.1	8.6	9.8
Yield y (kg)	4.5	5.9	7.0	7.8	7.2	6.8	4.5	2.7

(a) Calculate the product moment correlation coefficient for these data.
(b) Calculate Spearman's rank correlation coefficient for these data.
(c) Is there any evidence of a relationship between these variables? Justify your answer. (No formal test is required.)

AEB 1975

4. Three persons, P, Q and R, were asked to place in order of importance nine features of a house (A, B, C ... I). Calculate Spearman's rank correlation coefficients between the pairs of preferences as shown in the following table.

	A	B	C	D	E	F	G	H	I
P	1	2	4	8	9	7	6	3	5
Q	1	4	5	8	7	9	2	3	6
R	1	9	6	8	7	4	2	3	5

How far does this help to decide which pair from the three would be most likely to be able to compromise on a suitable house?

AEB 1975

5. The examination marks for a group of ten students in a Pure Mathematics and a Statistics paper are as shown.

Pure Mathematics	89	73	57	53	51	49	47	44	42	38
Statistics	51	53	49	50	48	21	46	19	40	43

Find the product moment correlation coefficient for the two sets of marks.

Place the marks in order of class position and calculate Spearman's rank correlation coefficient for the two papers.

The following is a quotation from a statistics text book. 'Rank correlation can be used to give a quick approximation to the product moment correlation coefficient.' Comment on this in the light of your results.

JMB

Chapter 18   Linear regression

18.1 Obtaining the Normal equations

If theoretical consideration or a high value of a correlation coefficient indicate that two variates are interdependent, we may wish to predict the value of one variate from a knowledge of the other. Consider, for example, the male population of England. Suppose X represents the weight and Y represents the height of the individuals of this population. In such a large population there will be a number of people of various heights for any (reasonable) weight. In particular, the height of the people of weight 70 kg will have a distribution as indicated in Fig. 18.1.

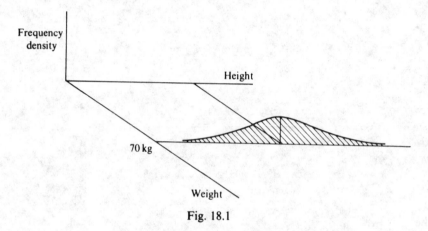

Fig. 18.1

Clearly if we wish to guess the height of a person chosen at random whose weight is 70 kg, the best estimate is the mean of this distribution. In general for any given value x of X the best prediction for the corresponding y would be the mean of the values of Y associated with this x. The line which passes through those mean values is called the line of regression of Y on X. A knowledge of this line would enable us to predict the value of Y associated with a particular value of X.

If the variates are jointly Normally distributed then it can be shown that this regression line will be straight. In practice if the variates have reasonably symmetrical distributions, the line of regression is assumed, to be straight.

It is the concern of this chapter to find an approximation to this line. Consider, for instance, the scatter diagram representing the data given in the introduction to the previous chapter concerning the weight and height of a sample of people (Fig. 18.2).

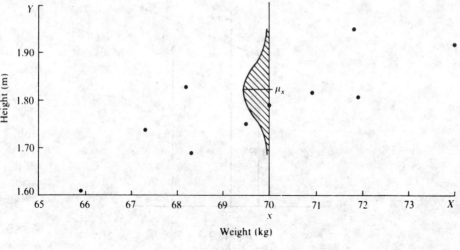

Fig. 18.2

The product moment correlation coefficient was found to be 0.8 (approx.), indicating a strong relationship. Attempting to estimate the line of regression of Y on X (the mean value of Y for given value of X) by eye is often quite adequate. This is particularly true of the experiments performed by scientists where the error due to measurements can be quite small; in the example above, however, it is rather more difficult. We shall seek an objective method of estimating the line of regression.

If we suppose the actual equation of the line of regression is

$$Y = \alpha + \beta X$$

then we wish to estimate the constants α and β using the data in the sample. Suppose that the estimated line is $Y = a + bX$, where a and b are the estimators of α and β respectively.

As it happens, there is in the sample a man whose weight is 70 kg and whose height is 1.79 m. As previously indicated, this height is itself a random value from an (approximately) Normal distribution, and we should expect this value to be quite close to the mean of this distribution (see Fig. 18.2). Since the actual line of regression passes through this mean value we would like this sample value to be quite close to our estimated line of regression.

The deviation will be

$$1.79 - (a + b \times 70) \qquad \text{(see Fig. 18.3)}.$$

Similarly for the point (x, y) in our sample, the deviation between the actual height and the (as yet unknown) predicted value is

$$d = y - (a + bx)$$
$$= y - a - bx$$

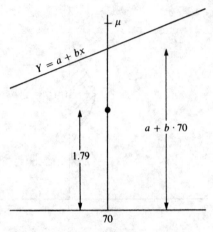

Fig. 18.3

Since we expect these deviations to be quite small, perhaps we ought to choose values for a and b which makes the sum of all these deviations (for the sample data) as small as possible. Unfortunately this does not work, since any line passing through the point $(\bar{x}, \bar{y})$ has a sum of deviations equal to zero! For suppose

$$\bar{y} = a + b\bar{x}, \quad \begin{array}{l}\text{ensuring that the line} \\ \text{passes through } (\bar{x}, \bar{y})\end{array}$$

$$\Leftrightarrow \quad \frac{\Sigma y}{n} = a + b\frac{\Sigma x}{n}$$

$$\Leftrightarrow \quad \Sigma y = na + b\Sigma x$$

$$\Leftrightarrow \quad \Sigma y = \Sigma(a + bx)$$

$$\Leftrightarrow \quad \Sigma[y - (a + bx)] = 0$$

$$\Leftrightarrow \quad \Sigma d = 0$$

Thus ensuring that the sum of the deviation is as small as possible does not help us to find values for a and b. We have met this problem before when we considered the standard deviation in Chapter 4. There we considered the sum of deviations from the mean and found this to be an unsatisfactory measure of spread. By analogy with the standard deviation we try to make the sum of the squares of the deviations as small as possible. In other words we attempt to minimise

$$D = \Sigma d^2 = \Sigma(y - a - bx)^2.$$

This is called the *method of least squares* and was first proposed by Gauss.

Derivation of Normal equations

Those familar with calculus will know that to minimise $D = \Sigma d^2$ we need to set the derivative to zero. In this case we think of a and b as the variables, the x and y values over which the summation takes place are fixed once the sample has been selected.

To minimise two unknowns we differentiate with respect to each variable in turn treating the other variable as a constant. The process is called partial differentiation and the notation used is standard.
Thus

$$\frac{\partial D}{\partial a} = \Sigma - 2(y - a - bx), \qquad \text{treating } b \text{ as a constant}$$

and

$$\frac{\partial D}{\partial b} = \Sigma - 2x(y - a - bx), \qquad \text{treating } a \text{ as a constant.}$$

Setting these equations to zero and simplifying we obtain

$$\Sigma y - na - b\Sigma x = 0$$

$$\Sigma xy - a\Sigma x - b\Sigma x^2 = 0$$

or alternatively

Definition 18.1

$$\begin{array}{ll} N_1 & \Sigma y = na + b\Sigma x \\ N_2 & \Sigma xy = a\Sigma x + b\Sigma x^2 \end{array} \left.\begin{array}{l} \\ \end{array}\right\} \quad \begin{array}{l} \text{the Normal} \\ \text{equations} \end{array}$$

The so called Normal equations are easy to remember by comparing with the original equation

$$Y = a + bX$$

Now, once the sample is known, then Σxy, Σx^2, Σx and Σy can be computed and substituted into the Normal equations. The resulting equations can be solved simultaneously to find our estimates a and b. (See Exercise 18.1 for practice.)

The analysis is now taken a stage further.

18.2 The equation of the line of regression of Y on X

If we divide equation N_1 by n we obtain

$$\bar{y} = a + b\bar{x}$$

an important result indicating that whatever our final choice of a and b the line will pass through the point $(\bar{x}, \bar{y})$. A knowledge of this fact helps when trying to draw the line of regression of Y on X by eye!
If we multiply N_1 by Σx and N_2 by n and then subtract we obtain

$$n\Sigma xy - \Sigma x\, \Sigma y = b(n\Sigma x^2 - (\Sigma x)^2)$$

On dividing both sides by n^2 this simplifies to

$$S_{xy} = bS_x^2,$$

where S_{xy} represents the covariance of the sample, giving the gradient of the line of regression as

$$b = \frac{S_{xy}}{S_x^2}$$

Since the line of regression of Y on X passes through $(\bar{x}, \bar{y})$, an alternative form of the line of regression of Y on X is

$$Y - \bar{y} = \frac{S_{xy}}{S_x^2}(X - \bar{x}) \qquad (1)$$

Alternative forms of the equation of the line of regression

The choice of which form of the equation we use depends on the way the data is presented to us. The following are equivalent to the above equation (1)

$$Y - \bar{y} = \frac{\Sigma(x - \bar{x})(y - \bar{y})}{\Sigma(x - \bar{x})^2} \ (X - \bar{x})$$

$$\Leftrightarrow \ Y - \bar{y} = \frac{\Sigma xy - \dfrac{\Sigma x \, \Sigma y}{n}}{\Sigma x^2 - \dfrac{(\Sigma x)^2}{n}} \ (X - \bar{x}) \qquad (2)$$

On dividing the original equation (1) by S_y we obtain

$$\frac{Y - \bar{y}}{S_y} = \frac{S_{xy}}{S_x S_y}\left(\frac{X - \bar{x}}{S_x}\right)$$

$$\Leftrightarrow \boxed{Z_y = r Z_x} \qquad (3)$$

where r is the product moment correlation coefficient.
The above is easy to remember!

Example 18.1

The data below represents the marks obtained by some students in two successive tests. A student absent from the second test scored 13 on the first test. Estimate the mark he might have obtained in the second test.

First test X	18	16	10	11	10	10	8	7
Second test Y	16	17	8	15	13	12	16	3

X	Y	X²	XY
18	16	324	288
16	17	256	272
10	8	100	80
11	15	121	165
10	13	100	130
10	12	100	120
8	16	64	128
7	3	49	21
$\Sigma x = 90$	$\Sigma y = 100$	$\Sigma x^2 = 1114$	$\Sigma xy = 1204$

$$\bar{y} = 12.5 \quad \bar{x} = 11.25$$

Now we obtain the line of regression

$$Y - \bar{y} = \frac{\Sigma xy - \dfrac{\Sigma x \, \Sigma y}{n}}{\Sigma x^2 - \dfrac{(\Sigma x)^2}{n}} \; (X - \bar{x})$$

so

$$Y - 12.5 = \frac{1204 - 1125}{1114 - 1012.5} \; (X - 11.25)$$

$$\Rightarrow \quad Y - 12.5 = 0.778 \, (X - 11.25)$$

$$\Rightarrow \quad Y = 3.744 + 0.778 \, X$$

The student scored 13 on his first test so assuming he is similar in ability to the other students we can substitute $X = 13$ into the line of regression of Y on X to obtain $Y \simeq$ 14 (actually $Y = 13.858$).

Suppose a further student missed the first test, but gained 15 marks in the second test. What is the best estimate in this case? The student may feel that the equation of the line of regression of Y on X can be used in this case, but as we shall see this is not usually so.

18.3 The line of regression of X on Y

Suppose a point C is chosen at random on the line AB 10 cm long. So the length X of AC is uniformly distributed from 0 to 10. Suppose further a point D is chosen, again at random on the line AC so that the length Y of AD is uniformly distributed from 0 to X.

The possible values of X and the corresponding Y values are shaded in Fig. 18.4.

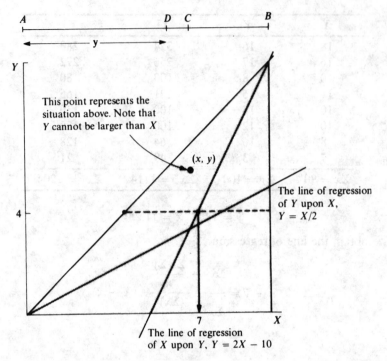

Fig. 18.4

The line of regression of Y on X passes through the mean of the possible Y values associated with each value of X. The equation of this line is

$$Y = \frac{X}{2}.$$

Thus, if X is 8 then the estimated Y is 4. If, however, Y is known to be 4, what is the best estimate of X? The mean value of the possible X values is 7 and this is our best estimate of X. In general we see that the line of regression of X on Y, the line passing through the mean value of X corresponding to each particular value of Y is

$$Y = 2X - 10.$$

It is generally true that there are two lines of regression, only when the correlation is perfect will the two lines be identical.

From the symmetry of the discussion above (Section 18.2) the equation of the line of regression of X on Y is

$$Z_x = r Z_y \quad \text{(see the corresponding equation (3) above)}$$

It can easily be seen that only if $r = \pm 1$ will the equations be identical. If $r = 0$ then

$$Z_x = 0 \quad \text{and from equation (3) } Z_y = 0$$

$$\Rightarrow \quad x = \bar{x} \quad \text{and} \qquad\qquad Y = \bar{y}$$

In other words when $r = 0$ a knowledge of X gives us no additional information concerning the value of Y and vice versa.

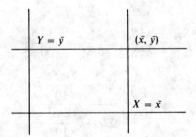

To find a suitable form of the line of regression of X on Y

$$Z_x = rZ_y$$

$$\Leftrightarrow \quad \frac{X - \bar{x}}{S_x} = \frac{S_{xy}}{S_x S_y} \frac{(Y - \bar{y})}{S_y}$$

$$\Leftrightarrow \quad X - \bar{x} = \frac{S_{xy}}{S_y^2} (Y - \bar{y})$$

$$\Leftrightarrow \quad \boxed{X - \bar{x} = \frac{\Sigma xy - \dfrac{\Sigma x \, \Sigma y}{n}}{\Sigma y^2 - \dfrac{(\Sigma y)^2}{n}} \; (Y - \bar{y})} \qquad (4)$$

18.4 The use of coding

Suppose $x = a + bu$ and $y = c + dv$ where, a, b, c and d are constants then as has been noted

$$S_{xy} = bd S_{uv}$$

$$S_y = d S_v$$

$$S_x = b S_u$$

$$\bar{x} = a + b\bar{u}$$

$$\bar{y} = c + d\bar{v}$$

The use of coding will be explained by way of an example.
Consider the data first introduced in Chapter 17 p. 249, concerning the height and weight of a small sample of people, reproduced below.

| Weight (kg) | 67.3 | 68.3 | 70.9 | 70.0 | 65.9 | 68.2 | 71.8 | 74.0 | 71.9 | 69.5 | X |
| Height (m) | 1.74 | 1.69 | 1.82 | 1.79 | 1.61 | 1.83 | 1.95 | 1.92 | 1.81 | 1.75 | Y |

We find the line of regression of Y on X as follows. Let $X = U + 70$ and $Y = 1/100 \, V + 1.80$. So with the above notation $a = 70, b = 1, c = 1.80$ and $d = 1/100$. Using the information given in the table on p. 254

$$\Sigma u = -2.2 \quad \Sigma v = -9 \quad \Sigma u^2 = 54.14 \quad \Sigma uv = 188.6$$

So

$$\frac{S_{xy}}{S_x^2} = \frac{bd\, S_{uv}}{b^2\, S_u^2} = \frac{d}{b}\frac{S_{uv}}{S_u^2} = \frac{1}{100}\frac{S_{uv}}{S_u^2}$$

$$= \frac{1}{100}\;\frac{\Sigma uv - \dfrac{\Sigma u\, \Sigma v}{10}}{\Sigma u^2 - \dfrac{(\Sigma u)^2}{10}}$$

$$= \frac{1}{100}\left(\frac{188.6 - 1.98}{54.14 - 0.484}\right)$$

$$= 0.035$$

and

$$\bar{x} = \bar{u} + 70 \qquad \bar{y} = \frac{1}{100}\bar{v} + 1.80$$

$$= 69.78 \qquad\qquad = 1.80 - 0.009$$

$$\qquad\qquad\qquad = 1.791.$$

substituting into

$$Y - \bar{y} = \frac{S_{xy}}{S_x^2}\;(X - \bar{x})$$

we have

$$Y - 1.791 = 0.035\;\;(X - 69.78)$$

that is

$$Y = -0.651 + 0.035X$$

This equation can now be used to estimate the height of a person for a given weight, either by direct calculation or by using the graph in Fig. 18.5.

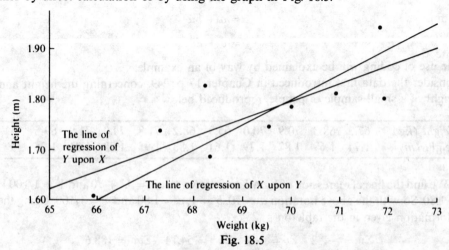

Fig. 18.5

The simplest way to draw this graph is to plot the point $(\bar{x}, \bar{y})$, i.e. (69.78, 1.791) and then to choose a convenient value of x (say $x = 65$) and calculate the corresponding value of y ($y = 1.624$). The line can then be drawn through these two points.

The line of regression of X on Y has also been drawn on the diagram in Fig. 18.5.

Further comments

The discussion so far has assumed that a sample is chosen randomly, that is it is representative of the population. The reliability of the estimated line of regression rests on this assumption. Thus it is pointless to try to use a line of regression to estimate the value of one variate when the value of the corresponding variate is not a member of the population from which the sample was taken. It would be most unwise to estimate the height of a Japanese gentleman using the line of regression of Y on X calculated above! Further one must proceed with caution when extending a line of regression beyond the limits of the sample. For example, the estimated height of a person whose weight is 18.6 kg using the line of regression above is zero!

We use the line of regression of Y on X when we wish to predict a value of Y from a value of X. Similarly we use the line of regression of X on Y when we wish to predict a value of X from a value of Y.

Example 18.2

Use the graph in Fig. 18.5 to estimate the weight X of a person whose height Y is 1.70 m.

Clearly we need to use the line of regression of X on Y which gives an estimated weight of 67.95 kg.

Had we used the other (and inappropriate) line of regression we would have obtained an estimated weight of 67.2 kg!

Sometimes, particularly in the sciences, one of the variables X (say) can be controlled, so that any variation can be neglected. In such circumstances we can use the line of regression of Y on X to predict Y values but we could not use a line of regression of X on Y to predict an X value since X is not a variate.

Example 18.3

The items in a sample each have associated with them values of the variates X and Y. Write down the formula for the sample product moment correlation coefficient between X and Y and the equation of the regression line X on Y.

State the conclusions you would draw if the correlation coefficient was (i) 0 (ii) -1.

For a sample of ten such items, the following data was established.

$$\Sigma X = 650 \quad \Sigma Y = 1430 \quad \Sigma Y^2 = 284108 \quad \Sigma XY = 104950 \quad \text{and} \quad \Sigma X^2 = 61200$$

Calculate the sample product moment correlation coefficient between X and Y.

Find, also, the equation of the regression line X on Y.

The product moment correlation coefficient r is given by

$$r = \frac{S_{xy}}{S_x S_y}$$

Where S_{xy} is the sample covariance $1/n \; \Sigma(x - \bar{x})(y - \bar{y})$ and S_x^2 and S_y^2 are the sample variances of X and Y respectively. The line of regression of X on Y is

$$X - \bar{x} = \frac{S_{xy}}{S_y^2} \; (Y - \bar{y}) \qquad ①$$

(i) If $r = 0$ then we deduce that there is no linear correlation. (This could mean that the variables are independent or that they are related in a non-linear way.)

(ii) If $r = -1$ then there is perfect negative correlation. Thus a high value of X corresponds with a low value of Y and a low value of X corresponds with a high value of Y.

$$S_{xy} = \frac{1}{n} \Sigma(x - \bar{x})(y - \bar{y})$$

$$= \frac{1}{n} \left(\Sigma xy - \frac{\Sigma x \, \Sigma y}{n} \right)$$

$$= \frac{1}{10} \left(104\,950 - \frac{650 \times 1430}{10} \right)$$

$$= 1200$$

$$S_x^2 = \frac{\Sigma x^2}{n} - \left(\frac{\Sigma x}{n} \right)^2$$

$$= 1895$$

$$S_y^2 = \frac{\Sigma y^2}{n} - \frac{(\Sigma y)^2}{n}$$

$$= 7961.8$$

$$r = \frac{S_{xy}}{S_x \, S_y}$$

$$= \frac{1200}{\sqrt{1895 \times 7961.8}}$$

$$= 0.309$$

Substituting into the formula ① for the line of regression of X on Y we have

$$X - 65 = \frac{1200}{7961.8} \; (Y - 143)$$

$$X = 0.151 Y + 43.45$$

Exercise 18.1

In this exercise reference is made to lines of regression of x on y using lower case letters. This follows common practice and should not lead to confusion. The distinction is retained in the text between upper and lower case letters to maintain consistency of use within this book.

1. Use the Normal equations to obtain the line of regression of y on x for the following set of data

x	3	5	7	4
y	2	9	8	5

Plot the line on a scatter diagram.

2. Draw a scatter diagram for the following data

x	4	9	6	4	9
y	8	3	7	7	4

Plot the point $(\bar{x}, \bar{y})$ and draw the line of 'best fit' by eye.
   Calculate the line of regression of Y on X and X on Y. Draw each of these lines on your scatter diagram. Use the appropriate line to estimate the value of x corresponding to a value of 6 for y.

3. If $\Sigma x = 35$, $\Sigma y = 65$, $\Sigma x^2 = 290$ and $\Sigma xy = 32$ for ten pairs of values of x and y find a line of regression.

4. If X and Y are related to U and V by

$$X = 10U + 2 \quad \text{and} \quad Y = 20V + 3,$$

where $\Sigma u = 2.3$, $\Sigma v = 1.9$, $\Sigma u^2 = 6.4$, $\Sigma v^2 = 5.1$ and $\Sigma uv = 3.8$ for eight pairs of values of U and V, find
(a) the product moment correlation coefficient,
(b) the line of regression of X on Y.

5. The table below shows five pairs of values of x and y.

x	-2	-1	0	1	2
y	30	22	24	18	14

Find
(a) the line of regression of y on x,
(b) if x and w are exactly related by $w = 3(x - 4)$, deduce the line of regression of y on w.

6. Use the information given on p. 254 to show that the line of regression of X on Y for the weights x and heights y of a sample of people first introduced in Chapter 17 is $x = 33.41 + 20.31y$.

7. The results of an experiment to determine how the percentage sand content of soil, y, varies with depth in cm below ground level x are given in the following table.

x	0	6	12	18	24	30	36	42	48
y	80.6	63.0	64.3	62.5	57.5	59.2	40.8	46.9	37.6

Calculate
(a) the covariance of x and y;
(b) the product moment correlation coefficient of x and y;
(c) the equation of the line of regression of y on x.
Explain briefly why the product moment correlation coefficient is preferable to the covariance as a measure of the association between x and y.

MEI

8. Explain briefly how the principle of 'least squares' may be used to estimate the coefficients in a linear regression equation given pairs of values of a variable y and a predetermined variable x.
In an agricultural experiment is is desired to investigate the relationship between the yield (y) and the amount of water supplied (x). Seven values of x were chosen, and for each value of x the corresponding value of y was measured. The results (in suitable units) were as follows.

Amount of water	(x)	12	18	24	30	36	42	48
Yield	(y)	5.27	5.68	6.25	7.21	8.02	8.71	8.42

Find the linear regression equation of y on x.
Estimate the yield when 54 units of water are supplied.

JMB

9. In a certain heathland region there is a large number of alder trees where the ground is marshy but very few where the ground is dry. The number x of alder trees and the ground moisture content y are found in each of 10 equal areas (which have been chosen to cover the range of x in all such areas). The following is a summary of the results of the survey:

$$\Sigma x = 500, \ \Sigma y = 300, \ \Sigma x^2 = 27\,818, \ \Sigma xy = 16\,837, \ \Sigma y^2 = 10\,462.$$

Find the equation of the regression line of y on x.
Estimate the ground moisture content in an area equal to one of the chosen areas which contains 60 alder trees.   o and c

10. In Utopia the wholesale food price index (x) and the retail food price index (y) over twelve years had the following values in order.

x	100	98	96	97	95	92	89	87	88	90	86	88
y	100	97	95	98	94	95	90	89	91	93	89	97

Find the equations of the regression lines (i) of y on x, (ii) of x on y. Calculate the product moment coefficient of correlation.

When the wholesale food price index is 84, what is the estimated value of the retail price index?

AEB 1975

11. A scientist, working in an agricultural research station, believes there is a relationship between the hardness of the shells of eggs laid by chickens and the amount of a certain food supplement put into the diet of the chickens. He selects ten chickens of the same breed and collects the following data:

Chicken	A	B	C	D	E	F	G	H	I	J
Amount of food supplement x (g)	7.0	9.8	11.6	17.5	7.6	8.2	12.4	17.5	9.5	19.5
Hardness of shells y	1.2	2.1	3.4	6.1	1.3	1.7	3.4	6.2	2.1	7.1

(Hardness is measured on a 0–10 scale, 10 being the hardest. There are no units attached.)

(a) Calculate the equation of the regression line of y on x.

(b) Calculate the product-moment correlation coefficient.

(c) Do you believe that this linear model will continue to be appropriate no matter how large or small x becomes? Justify your reply.

AEB 1978

12. Ten boys compete in throwing a cricket ball, and the following table shows the height of each boy (x cm) to the nearest cm and the distance (y m) to which he can throw the ball.

Boy	A	B	C	D	E	F	G	H	I	J
x	122	124	133	138	144	156	158	161	164	168
y	41	38	52	56	29	54	59	61	63	67

Find the equations of the regression lines of y on x, and of x on y. No diagram is needed. Calculate also the coefficient of correlation.

Estimate the distance to which a cricket ball can be thrown by a boy 150 cm in height.

13. The body and heart masses of fourteen 10-month-old male mice are tabulated below:

Body mass (x) (g)	27	30	37	38	32	36	32
Heart mass (y) (mg)	118	136	156	150	140	155	157

Body mass (x) (g)	32	38	42	36	44	33	38
Heart mass (y) (mg)	114	144	159	149	170	131	160

(a) Draw a scatter diagram of these data.

(b) Calculate the equation of the regression line of y on x and draw this line on the scatter diagram.

(c) Calculate the product moment coefficient of correlation.

AEB 1977

14. Ten experiments result in ten pairs of measurements (X_i, Y_i) of two variables (X, Y) and $\Sigma X_i = \Sigma Y_i = 0$. It is desired to estimate the value of Y for a given value of X. In what circumstances would you consider it sensible to use an equation of the form $Y = mX$ for this purpose? Assuming these circumstances hold, explain with reasons how the constant m may be chosen and derive an expression for it in terms of (X_i, Y_i).

Calculate m for the observations given in the following table. (You should assume that $\Sigma X_i = \Sigma Y_i = 0$ without calculation.)

X_i	-1.5	-1.2	-0.7	-0.5	-0.1
Y_i	-5.6	-2.5	-2.2	-3.5	-1.1
X_i	0.2	0.4	0.7	1.1	1.6
Y_i	1.3	2.8	2.0	3.8	5.0

15. A large group of students took a trial exam in January and their O level the following July. A random sample of the marks obtained by these students is set out below.

Trial Mark	52	64	57	84	48	40	59	44	48
O Level Mark	46	55	56	70	42	36	47	26	38

Obtain the line of regression of the O Level marks on the trial marks. Draw this line of regression on the scatter graph of the above data. Estimate the O level mark obtained by a student whose trial mark was

(a) 40,

(b) 20.

Explain why you have more confidence in the accuracy of one of the estimates compared with the other.

16. The pressure p and volume V of a fixed mass of gas are believed to be related by an equation of the form

$$pV^\gamma = \text{constant},$$

where γ is a constant. From this equation obtain a linear relation between $x = \ln p$ and $y = \ln V$.

In 6 experiments on the fixed mass of gas, in each of which p was controlled and V measured, the results satisfied:

$$\Sigma x_i = 2.420, \quad \Sigma y_i = -1.708, \quad \Sigma(x_i^2) = 3.171,$$

$$\Sigma(y_i^2) = 1.561, \quad \Sigma x_i y_i = -2.224.$$

Here x_i and y_i are the values of $\ln p$ and $\ln V$ in the ith experiment, and p was measured in Nm^{-2} and V in m^3.

Calculate the equation of the line of regression of y on x. Hence estimate to 2 decimal places

(i) the value of γ;
(ii) the value of V when $p = 0.75\,Nm^{-2}$.

How would knowledge of the range of values of p in the experiment affect your confidence in your answer to (ii)?

MEI

17. Corresponding observations of the variables x and y and computed values of $(z = \log_{10} y)$ are shown in the following table.

x	5	6	7	8	9
y	12	14	15	15	18
z	1.079	1.146	1.176	1.176	1.255

x	10	11	12	13	14
y	16	20	19	19	24
x	1.204	1.301	1.279	1.279	1.380

$[\Sigma z = 12.275, \quad \Sigma xz = 118.852.]$

Plot y against x and z against x on separate diagrams. Describe the conditions for the valid use of the regression line of y on x and give a reason why it would be unwise to use it on the data but would probably be valid for z on x.

Obtain the equation of the regression line for z on x and hence show that the estimated relation connecting x and y is

$$y = 9.32 \times 1.064^x.$$

CAMB

18. Show that, if the regression line of y on x is derived from 7 pairs of values (x_1, y_1), $(x_2, y_2), \ldots, (x_7, y_7)$, where $x_i = a + id$, the estimated slope of the line may be written

$$\sum_{i=1}^{7} \frac{(i - 4)\,y_i}{28d}.$$

Hence, or otherwise, find the regression line of heights on age from the data:

Age (years)	5	6	7	8	9	10	11
Height (cm)	99	105	111	117	123	128	133

19. The following data represent the lengths (x) and breadths (y) of 12 cuckoos' eggs measured in millimetres.

x	22.3	23.6	24.2	22.6	22.3	22.3	22.1	23.3	22.2	22.2	21.8	23.2
y	16.5	17.1	17.3	17.0	16.8	16.4	17.2	16.8	16.7	16.2	16.6	16.4

Draw a scatter diagram for the data.

Obtain the least squares regression line of y on x and plot this on the scatter diagram.

JMB

20. It is thought that a straight line of the form $y = \beta x$ might serve as a model providing a rough estimate of the actual distance y along the streets of a city between two points whose direct distance apart is x. Show that for the pairs of observed values $(x_i, y_i), i = 1, 2, \ldots, n$, the regression line of y on x of this form gives the least squares estimate of β as

$$b = \frac{\sum\limits_{i=1}^{n} x_i y_i}{\sum\limits_{i=1}^{n} x_i^2}$$

The following is such a set of data.

Direct distance, x (km)	0.5	1.0	1.5	2.0	2.5	3.0	3.5	4.0	4.5	5.0
Actual distance along streets, y (km)	0.9	1.2	1.8	2.9	3.1	3.5	3.9	5.6	5.8	6.6

Draw a scatter diagram for the data.

Fit a line of the form $y = bx$ to the data and use it to estimate the actual distance along the streets between two points whose linear distance apart is 4.2 kilometres.

JMB

Chapter 19 Probability density functions (revisited)

19.1 An Example

A continuous variate X has the *exponential distribution* if its pdf is of the form

$$f(x) = \begin{cases} \lambda e^{-\lambda x}, & \text{for } x \geq 0, \\ 0 & , \quad \text{otherwise,} \end{cases}$$

for some constant $\lambda > 0$.

Check that this is a pdf and use integration by parts to obtain the mean and variance of X in terms of λ.

To show that the function above represents a pdf.

(i) $\lambda e^{-\lambda x} \geq 0$ for all x since $\lambda > 0$

(ii) $\lambda \int_0^\infty e^{-\lambda x}\, dx = \lambda \left[\frac{-e^{-\lambda x}}{\lambda} \right]_0^\infty$

$$= 1 \text{ as required}$$

$$E(X) = \lambda \int_0^\infty x e^{-\lambda x}\, dx$$

$$= \lambda \left[\frac{-x e^{-\lambda x}}{\lambda} \right]_0^\infty - \lambda \int_0^\infty -\frac{e^{-\lambda x}}{\lambda}\, dx$$

$$= 0 + \int_0^\infty e^{-\lambda x}\, dx$$

$$= \left[-\frac{e^{-\lambda x}}{\lambda} \right]_0^\infty$$

$$= \frac{1}{\lambda} \tag{1}$$

$$E(X^2) = \lambda \int_0^\infty x^2 e^{-\lambda x}\, dx$$

$$= \lambda \left[\frac{-x^2 e^{-\lambda x}}{\lambda} \right]_0^\infty - 2\lambda \int \frac{-x e^{-\lambda x}}{\lambda}\, dx$$

$$= 0 + \frac{2}{\lambda^2} \text{ by (1).}$$

Thus

$$\text{Var}(X) = E(X^2) - E(X)^2$$

$$= \frac{2}{\lambda^2} - \frac{1}{\lambda^2}$$

$$= \frac{1}{\lambda^2}.$$

Exercise 19.1

1. The variate X has a pdf given by

$$f(x) = \begin{cases} A \sin x, & 0 < x < \pi, \\ 0, & x \leqslant 0 \text{ or } x \geqslant \pi. \end{cases}$$

Find

(a) $P(X < \pi/3)$
(b) $E(X)$
(c) $\text{Var}(X)$

2. The variate X has a pdf of this form

$$f(x) = \frac{k}{1 + x^2}, \quad -\infty \leqslant x \leqslant \infty.$$

Find the value of k and hence find the probability that each of two independent observations is greater than 1.

3. The mileage (in thousands of miles) obtained from a certain brand of car tyre is a variate with pdf

$$f(x) = \begin{cases} k e^{-x/20}, & \text{for } x > 0, \\ 0, & \text{for } x \leqslant 0. \end{cases}$$

Find the value of k and the mean life of a tyre.
Find also the probability that a tyre will last
(a) at most 1000 miles,
(b) between 5 and 15 thousand miles.

4. A variate X is defined for all values and has a pdf given by

$$f(x) = \begin{cases} \dfrac{2}{x^2}, & \text{for } 1 \leqslant x \leqslant 2, \\ 0, & \text{otherwise.} \end{cases}$$

Find

(a) the mean of X,   (d) the median,
(b) the variance of X (e) $P(X < 5 \text{ or } X \geqslant 7)$.
(c) $P(X$ is larger than the mean),

5. The standard Normal distribution $N(0, 1)$ has a pdf given by

$$\varphi(z) = \frac{1}{\sqrt{2\pi}} e^{-z^2/2}, \quad -\infty < z < \infty.$$

Show by integration that the mean is zero and the variance is 1 for this distribution.

6. A normal variate X has a pdf of the form

$$\frac{1}{\sigma\sqrt{2\pi}} e\left[-\tfrac{1}{2}\left(\tfrac{x-\mu}{\sigma}\right)^2\right], \quad -\infty < x < \infty,$$

where X has mean μ and variance σ^2

(i) Write down an integral which represents $P(X < a)$.
(ii) By means of the substitution $z = (x - \mu/\sigma)$ show that the above probability is the same as $P(z < (a - \mu)/\sigma)$. Comment on the importance of this result.

7. (a) Determine $E(x)$ and $E(x^2)$, where $E(A)$ denotes the expected value of A, and hence the mean and variance of each of the distributions
   (i) the uniform (i.e. rectangular) distribution on the interval $a \leqslant x \leqslant b$,
   (ii) the uniform distribution on the integers $1, 2, \ldots, N$ given by

$$p(x) = \begin{cases} \dfrac{1}{N} & (x = 1, 2, \ldots, N) \\ 0 & \text{otherwise.} \end{cases}$$

   [You may assume that the sum of the squares of the first n positive integers is $\tfrac{1}{6}n(n + 1)(2n + 1)$.]
   (b) Given that x has a continuous distribution, show that $E\{(x - t)^2\}$ is a minimum when $t = \mu$, where μ is the mean of the distribution.

MEI

8. A variable X has a probability density function given by

$$0 \text{ for } x < a$$
$$f(x) \text{ for } a \leqslant x \leqslant \beta$$
$$0 \text{ for } x > \beta.$$

(x represents an actual value taken by the random variable X.)
   Give two conditions that must be satisfied by the function $f(x)$.
   In the case where $a = 0$, $\beta = 1$ and $f(x) = kx(x - 1)^2$, determine the value of k. Show that there is no suitable value for k when $a = -1$ and $\beta = 1$.
   Calculate the mean and variance of X for the distribution for which k has been determined. Find the value of X for which the probability density function is a maximum and sketch the probability density function.

JMB

9. Suppose that X is uniformly distributed over the interval $-1 \leqslant x \leqslant 1$. Find

(a) $P(x > \frac{3}{4} \mid |x| > \frac{1}{2})$

(b) $P(X^2 \leqslant \frac{1}{4})$

10. Potatoes are packed into sacks by a farmer. The weights of the sacks are distributed Normally with mean 50 kg and standard deviation 8 kg. Write down the probability density function of the weight of sacks. Determine the probability that the weight of a sack is greater than 60 kg.

Only those sacks whose weights are greater than 45 kg are sent directly to the distributor, the others being rejected and sent back for repacking. Show that the probability density function of the weight of sacks which are initially accepted is $f(x)$, where

$$f(x) = \begin{cases} 0, & x \leqslant 45; \\ \dfrac{A}{(128\pi)^{1/2}} \exp\left[-\dfrac{(x-50)^2}{128} \right], & x > 45; \end{cases}$$

and A is a constant. Determine the proportion of sacks which is accepted and hence find the value of A correct to two decimal places.

What is the probability that a sack which is accepted has a weight greater than 60 kg?

MEI

11. The pdf of X is given by

$$f(x) = 2e^{-2x}, \text{ for } x \geqslant 0.$$

Find the probability that a value of X is greater than 2.

If values of X are established one by one find the probability that the first value larger than 2 occurs as the nth such selection.

12. The random variable X has a Normal distribution. Its probability density function $f(x)$ is given by

$$f(x) = \frac{1}{\sigma\sqrt{2n}} \exp\left\{ -\frac{1}{2}\left(\frac{x-\mu}{\sigma}\right)^2 \right\} \quad \begin{array}{l} -\infty < x < \infty \\ -\infty < \mu < \infty, \quad \sigma > 0 \end{array}$$

Derive the mean and variance of this distribution. (You may assume that

$$\frac{1}{\sqrt{2}} \int_{-\infty}^{\infty} \exp(-t^2/2)dt = 1; \quad \exp(y) \equiv e^y)$$

Suppose $X_1 \ldots, X_n$ is a random sample from a Normal distribution which has mean and standard deviation 2. What is the distribution of the sample mean $\bar{X}$? How large should n be so that the probability that $|\bar{X} - \mu| \leqslant 0.5$ is at least 0.95? Comment on this result.

AEB 1977

13. A discrete distribution is defined by

$$P(X = r) = \lambda r, \quad (r = 1, 2, \ldots, n).$$

Find the value of λ and the mean of the distribution. Show that the variance is $(n - 1)(n + 2)/18$. What is the median value when $n = 12$?

14. The probability density function of x is given by

$$f(x) = \begin{cases} k(ax - x^2), & 0 \leqslant x \leqslant 2 \\ 0, & x < 0, \quad x > 2, \end{cases}$$

where k and a are positive constants.

Show that $a \geqslant 2$ and that $k = 3/(6a - 8)$.

Given that the mean value of x is 1, calculate the values of a and k. For these values of a and k sketch the graph of the probability density function and find the variance of x.

JMB

19.2 The distribution of functions of a uniform variate

Example 19.2

Suppose values of a continuous variate X uniformly distributed between 1 cm and 3 cm are used as radii of circles with area Y cm

Find

(a) the range of possible values of Y,
(b) the expected value of Y,
(c) the variance of Y,
(d) the probability that $Y < 2\pi$.

We have $Y = \pi X^2$ where X has pdf

$$f(x) = \begin{cases} \dfrac{1}{2}, & \text{for } 1 \leqslant x \leqslant 3, \\ 0, & \text{otherwise.} \end{cases}$$

(a) Clearly Y takes values between π and 9π.

(b)
$$\begin{aligned}
E(Y) &= E(\pi X^2) \\
&= \int_1^3 \pi x^2 \frac{1}{2} \, dx \quad \text{by definition} \\
&= \frac{\pi}{2} \left[\frac{x^3}{3} \right]_1^3 \\
&= \frac{13\pi}{3}
\end{aligned}$$

(c) To find Var (Y) we first compute

$$E(Y^2) = E(\pi^2 X^4)$$

$$= \frac{\pi^2}{2} \int_1^3 x^4 \, dx$$

$$= \frac{\pi^2}{2} \left[\frac{x^5}{5} \right]_1^3$$

$$= \frac{121\pi^2}{5}$$

Thus

$$\text{Var } (Y) = E(Y^2) - E(Y)^2$$

$$= \frac{121}{5}\pi^2 - \frac{169}{9}\pi^2$$

$$= \frac{244}{45}\pi^2$$

$$\simeq 53.5$$

(d) $P(Y < 2\pi) = P(\pi X^2 < 2\pi)$

$$= P(X^2 < 2)$$

$$= P(1 < X < \sqrt{2}), \quad \text{because of the restrictions on the values of } X,$$

$$= \frac{1}{2}(\sqrt{2} - 1)$$

Note that, in the above example, questions concerning the area Y could be reduced to ones concerning the radius X whose pdf was known.

We shall now be concerned with the problem of finding the pdf of Y.

Suppose X is a random variable, we shall use the notation $f_X(x)$ to represent the pdf of X.

For a continuous distribution we know that

$$P(X = a) = 0,$$

but

$$P(X \text{ is close to } a) = P\left(a - \frac{\delta x}{2} \leqslant X \leqslant a + \frac{\delta x}{2}\right)$$

$$= P(X \simeq a)$$

$$\simeq f_X(a)\delta x, \quad \text{as suggested in Fig. 19.1.}$$

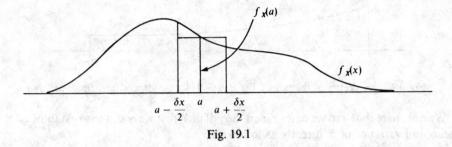

Fig. 19.1

We return now to the problem of finding the pdf of Y where $Y = \pi X^2$ and

$$f_X(x) = \begin{cases} \dfrac{1}{2}, & \text{for } 1 \leqslant x \leqslant 3, \\ 0, & \text{otherwise.} \end{cases}$$

We have

$$f_Y(y)\delta y \simeq P(Y \simeq y)$$

$$\simeq P(\pi X^2 \simeq y)$$

$$\simeq P\left(X \simeq \sqrt{\dfrac{y}{\pi}}\right), \quad \text{we ignore the negative square root since } X \geqslant 1,$$

$$\simeq f_X\left(\sqrt{\dfrac{y}{\pi}}\right)\delta x$$

$$= \begin{cases} \dfrac{1}{2}\delta x, & \text{if } 1 \leqslant \sqrt{\dfrac{y}{\pi}} \leqslant 3, \text{ that is } \pi \leqslant y \leqslant 9\pi, \\ 0, & \text{otherwise.} \end{cases}$$

Now assuming $\pi \leqslant y \leqslant 9\pi$

$$\dfrac{1}{2}\delta x \simeq \dfrac{1}{2}\left|\dfrac{\mathrm{d}x}{\mathrm{d}y}\right||\delta y| \quad \begin{array}{l}\text{Since } \mathrm{d}y/\mathrm{d}x \simeq \delta y/\delta x\text{; the modulus sign} \\ \text{ensures that the expression is positive} \\ \text{since } \delta x \text{ is positive.}\end{array}$$

$$\simeq \dfrac{1}{2}\cdot\dfrac{1}{2\pi x}\delta y, \quad \begin{array}{l}\text{since } y = \pi x^2, \\ \text{so } \dfrac{\mathrm{d}y}{\mathrm{d}x} = 2\pi x,\end{array}$$

$$\simeq \dfrac{1}{4\sqrt{\pi y}}\delta y$$

We see then that the pdf of Y is

$$f_Y(y) = \begin{cases} \dfrac{1}{4\sqrt{\pi y}}, & \text{for } \pi \leqslant y \leqslant 9\pi, \\ 0, & \text{otherwise.} \end{cases}$$

Fig. 19.2   A sketch of the pdf of $Y . f_Y(y) = 1/4\sqrt{\pi y}$ for $\pi \leqslant y \leqslant 9\pi$, 0 otherwise.

We note here that had we determined the pdf of Y first we could have established the mean and variance of Y directly as follows

$$E(Y) = \int_{\pi}^{9\pi} y \cdot \frac{1}{4\sqrt{\pi y}} \, dy$$

$$= \frac{1}{4\sqrt{\pi}} \int_{\pi}^{9\pi} y^{1/2} \, dy$$

$$= \frac{1}{4\sqrt{\pi}} \left[\frac{2y^{3/2}}{3} \right]_{\pi}^{9\pi}$$

$$= \frac{13\pi}{3}.$$

Similarly we can find the variance of Y though the reader will probably prefer the method used earlier.

To find the pdf of $Y = g(X)$ when the pdf of X is known

Suppose X and Y are variates such that

$$Y = g(X),$$

where for each value of Y there corresponds just one value of X.

Suppose further that X is rectangularly distributed between a and b where $a < b$.

We restrict our attention to those values of X for which $a \leqslant X \leqslant b$ and thus Y lies between $g(a)$ and $g(b)$.

$$P(Y \simeq y) \simeq P(X \simeq x), \quad \text{where } y = g(x)$$

and thus

$$f_Y(y)|\delta y| \simeq \frac{1}{b-a} \, \delta x \quad \text{(it is possible that } \delta y < 0\text{)}.$$

So

$$f_Y(y) \simeq \frac{1}{b-a} \left| \frac{\delta x}{\delta y} \right|$$

and in the limit

$$f_Y(y) = \begin{cases} \dfrac{1}{b-a}\left|\dfrac{dx}{dy}\right|, & \text{within the range } g(a) \text{ to } g(b), \\ 0, & \text{otherwise,} \end{cases}$$

where the function is written in terms of y.

We check this result by considering the earlier example where

$$Y = \pi X^2 \quad \text{and} \quad 1 \leqslant X \leqslant 3$$

thus

$$\frac{dY}{dX} = 2\pi X$$

$$= 2\pi \sqrt{\frac{Y}{\pi}}$$

$$= 2\sqrt{\pi Y}$$

So

$$f_Y(y) = \begin{cases} \dfrac{1}{2}\cdot\dfrac{1}{2\sqrt{\pi y}} = \dfrac{1}{4\sqrt{\pi y}}, & \text{if } \pi \leqslant y \leqslant 9\pi, \\ 0, & \text{otherwise.} \end{cases}$$

Example 19.3

Spheres are produced with volume V and radius R where R is rectangular distributed between 0.9 cm and 1.1 cm. Find the pdf of V and $E(V)$.

Since $V = 4/3\pi R^3$, the values of V lie between $4/3\pi\,0.9^3$ and $4/3\pi\,1.1^3$, that is between 3.05 and 5.58 approximately.

Within these limits

$$f_V(v) = \frac{1}{0.2}\cdot\frac{dr}{dv} \quad \text{and since} \quad v = \frac{4}{3}\pi r^3$$

$$= \frac{5}{4\pi}\left(\frac{4\pi}{3v}\right)^{2/3} \qquad \frac{dv}{dr} = 4\pi r^2$$

$$f_V(v) = 0 \text{ elsewhere.} \qquad\qquad = 4\pi\left(\frac{3v}{4\pi}\right)^{2/3}$$

To find $E(V)$ we can either use the pdf of V or continue as follows:

$$E(V) = E\left(\frac{4}{3}\pi R^3\right)$$

$$= \frac{4\pi}{3} E(R^3)$$

$$= \frac{4\pi}{3} \int_{0.9}^{1.1} r^3 \cdot 5\, dr$$

$$= \frac{5\pi}{3} [r^4]_{0.9}^{1.1}$$

$$\simeq 4.23$$

Example 19.4

A sector of a circle of unit radius and angle Θ is cut from a sheet of metal and bent to form the curved surface of a right circular cone of height X. If the angle Θ is uniformly distributed between 0 and 2π obtain the pdf of X.

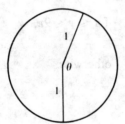 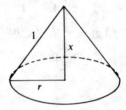

For any particular θ let r represent the radius of the resulting cone. We need to obtain a relationship between x and θ.

Now

$$r = \frac{\theta}{2\pi}$$

and so

$$x^2 + \left(\frac{\theta}{2\pi}\right)^2 = 1.$$

Differentiating with respect to x

$$2x + \frac{\theta}{2\pi^2} \frac{d\theta}{dx} = 0$$

$$\Rightarrow \quad \frac{d\theta}{dx} = \frac{-4\pi^2 x}{\theta}$$

$$= \frac{-2\pi x}{\sqrt{1-x^2}}, \quad \text{since } \theta = 2\pi\sqrt{1-x^2}.$$

The pdf of X is given by

$$f_X(x) = \frac{1}{2\pi} \cdot \frac{2\pi x}{\sqrt{1-x^2}} \quad \text{ignoring the sign}$$

$$= \frac{x}{\sqrt{1-x^2}}, \quad \begin{array}{l}\text{for values such that } 0 < \theta \leqslant 2\pi, \\ \text{that is } 0 \leqslant x \leqslant 1,\end{array}$$

$$= 0 \quad \text{otherwise.}$$

19.3 The cumulative distribution function

The (cumulative) distribution function, $F_X(x)$, of a variate X is defined for *all real values* x by

$$F_X(x) = P(X \leqslant x)$$

Thus if X has pdf $f_X(x)$ then

$$F_X(x) = \int_{-\infty}^{x} f_X(t)\,dt \quad (t \text{ is used here as a dummy variable})$$

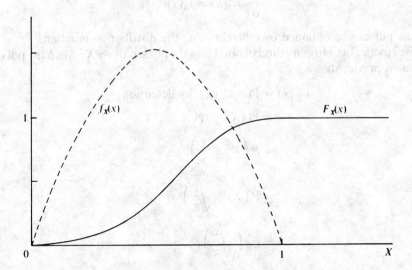

Fig. 19.3   A sketch of a pdf together with its associated (cumulative) distribution function.

If X is a discrete variate defined for values $x = 1, 2, 3, 4, \ldots, n$, then

$$F_X(x) = 0, \quad \text{for } x < 1,$$

$$= \sum_{1 \leqslant r \leqslant x} P(X = r), \quad \begin{array}{l}\text{for } 1 \leqslant x \leqslant n \\ \text{and the summation is} \\ \text{taken over all the integers} \\ r \text{ between 1 and } x,\end{array}$$

$$= 1, \quad \text{for } x > n.$$

The relationship between the distribution function and the CF polygon

The reader will, no doubt, recall drawing cumulative frequency polygons for examples of data. If the data was a random sample then the CF polygon can be thought of as an approximation to the distribution function which in the case of continuous data will be a smooth curve.

Some writers prefer to improve the approximation by drawing a cumulative frequency curve. This is produced in exactly the same way as the CF polygon except that instead of joining the plotted points by line segments, a smooth curve is drawn through the points.

The relationship between the distribution function and the pdf

We have noted already that

$$F_X(x) = \int_{-\infty}^{x} f_X(t)\, dt$$

A moment's thought will convince one that

$$\frac{d}{dx} F_X(x) = f_X(x)$$

Thus the pdf can be obtained by differentiating the distribution function.

This gives us an alternative method of obtaining the pdf for $Y = \pi X^2$ when the pdf of X is as stated previously.

$$F_Y(y) = P(Y \leqslant y), \quad \text{by definition};$$

$$= P(\pi X^2 \leqslant y)$$

$$= P\left(X^2 \leqslant \frac{y}{\pi}\right)$$

$$= P\left(X \leqslant \sqrt{\frac{y}{\pi}}\right)$$

$$= F_X\left(\sqrt{\frac{y}{\pi}}\right)$$

$$= \frac{1}{2}\left(\sqrt{\frac{y}{\pi}} - 1\right), \quad \text{if } \pi \leqslant y \leqslant 9\pi.$$

Now, differentiating with respect to y, we have

$$f_Y(y) = \frac{1}{4\pi}\left(\frac{y}{\pi}\right)^{-1/2}$$

$$= \frac{1}{4\sqrt{\pi y}}, \quad \text{if } \pi \leqslant y \leqslant 9\pi,$$

$$= 0, \quad \text{otherwise}; \quad \text{since outside the above range } F_Y(y) \text{ is constant.}$$

Exercise 19.2

1. The length X of the side of a square is uniformly distributed between 1 and 2. Show that the area Y is distributed between 1 and 4 with pdf

$$f(y) = \frac{1}{2} y^{-1/2}$$

2. The length X of the side of a cube is uniformly distributed between 1 and 2. Find the pdf of V, the volume of the cube. Sketch the pdf of V and find the mean and variance of V.

3. The radius X of a circle is uniformly distributed between 1 and 4. Find the pdf of
   (a) the circumference Y of the circle,
   (b) the area Z of the circle.

4. From a point on the circumference of a circle of radius a, a chord is drawn in a random direction so that if θ is the angle between the chord and the diameter through the point, then θ has a Rectangular distribution between $-\frac{\pi}{2}$ and $\frac{\pi}{2}$. Find the expected value and variance of the length of the chord. Show that the probability is 1/3 that the length of the chord exceeds the length of the side of an equilateral triangle inscribed in the circle.

MEI

5. A is a fixed point on a circle of centre O and radius r, and P is a point chosen at random on the circle, so that the angle at O in the triangle AOP is uniformly distributed between 0 and π.
   Show that the probability density function of the length X of the chord AP is

$$\frac{2}{\pi \sqrt{4r^2 - x^2}}$$

   for x between 0 and $2r$, and zero outside this range. (x represents an actual value taken by the random variable X.)
   Calculate the probability that the chord is longer than the radius. Find also the median value of X.
   Q is a fixed point on the circle and $AQ = k$. Find in terms of k and r the probability that $AP < AQ$.

JMB

6. A rectangle of area A square metres has a perimeter of 20 metres and a side is of length x metres, where x is uniformly distributed between 0 and 2. Show that the probability density function of A is

$$\frac{1}{4\sqrt{(25 - A)}}, \quad (0 \leqslant A \leqslant 16).$$

   Find the mean and variance of A.

JMB

7. A continuous random variable X takes values between 0 and 1 only and its probability density function f is given by $f(x) = kx^2(1 - x)$. Prove that $k = 12$, and find $E(X)$.

   The cumulative distribution function F is such that $F(x) = P(X \leqslant x)$. Express $F(x)$ in terms of x, and sketch the graph of $y = F(x)$. Verify by calculation that the median of X lies between 0.60 and 0.62.

CAMB

8. A child rides on a roundabout and his father waits for him at the point where he started. His journey may be regarded as a circular route of radius six metres and the father's position as a fixed point on the circle. When the roundabout stops, the shorter distance of the child from the father, measured along the circular path, is S metres. All points on the circle are equally likely stopping points so that S is uniformly distributed between 0 and 6π. Find the mean and variance of S.

   The direct linear distance of the child's stopping point from the father is D metres. Show that the probability density function of D is

$$\frac{2}{\pi\sqrt{(144 - D^2)}},$$

for D between 0 and 12 and zero outside this range.

   The father's voice can be heard at a distance of up to 10 m. Find to two decimal places the probability that the child can hear his father shout to him when the roundabout stops.

JMB

9. A measuring flask is so shaped that the volume V of the liquid in it is related to the height h of the surface of the liquid above the base of the flask by $V = \pi h^3/9$. The flask is used to measure volumes V of liquid which are uniformly distributed between 0 and 1 units. Find the probability density function of the distribution of the corresponding values of h and sketch the graph of this function. Calculate

   (a) the probability that h is less than 1 unit,
   (b) the mean of h,
   (c) the variance of h,
   (d) the probability that h is less than the mean of h,

$$\left(\text{Take } \sqrt[3]{\frac{9}{\pi}} \text{ as } 1.42\right).$$

JMB

10. The time t seconds between the arrival of successive vehicles at a point on a country road has a probability density function $ke^{-\lambda t}$, $t \geqslant 0$, where k and λ are positive constants. Find k in terms of λ and sketch the graph of the probability density function.

    Given that $\lambda = 0.01$, find the mean and variance of t.

$$\left[\text{You may use the result } \int_0^\infty t^r e^{-\lambda t}\, dt = \frac{r!}{\lambda^{r+1}}, r = 0, 1, 2, \ldots\right]$$

An elderly pedestrian takes 50 seconds to cross the road at this point. With $\lambda = 0.01$, calculate the probability that, if he sets off as one vehicle passes, he will complete the crossing before the next vehicle arrives. Calculate also the probability that, if he adopts the same procedure on the return journey, he completes each crossing without a vehicle arriving while he is doing so.

JMB

11. Three points are taken at random and independently on a circle (so that the probability that one of the points lies on a given element of the circle is proportional to the length of the element but independent of its position). Show that the probability that they lie on the same semicircle is 3/4. Show also that the probability that the three points form a triangle each of whose angles is acute is 1/4.

MEI

12. A variate X is uniformly distributed between 0 and a. Write down the value of $E(X)$ and determine the standard deviation of X.

Two independent observations X_1 and X_2 are obtained from X. By treating (X_1, X_2) as Cartesian co-ordinates and considering the lines $X_1 + X_2 = Y$ show that

$$P(X_1 + X_2 \leqslant Y) = \frac{Y^2}{2a^2}, \quad \text{if} \quad 0 \leqslant Y \leqslant a,$$

and find its value if $a < Y \leqslant 2a$. Hence find the pdf of $Y = X_1 + X_2$.

13. The continuous random variable X has probability density function $f(x)$ defined by

$$f(x) = \begin{cases} \dfrac{c}{x^4} & (x < -1) \\ c(2 - x^2) & (-1 \leqslant x \leqslant 1) \\ \dfrac{c}{x^4} & (x > 1). \end{cases}$$

(a) Show that $c = \frac{1}{4}$.
(b) Sketch the graph of $f(x)$.
(c) Determine the cumulative distribution function $F(x)$.
(d) Determine the expected value of X and the variance of X.

CAMB

14. A random variable X has (cumulative) distribution function

$$F(x) = \begin{cases} 0 & (x \leqslant 0) \\ \frac{1}{4}x^2 & (0 \leqslant x \leqslant 1) \\ \alpha x + \beta & (1 \leqslant x \leqslant 2) \\ \frac{1}{4}(5 - x)(x - 1) & (2 \leqslant x \leqslant 3) \\ 1 & (3 \leqslant x). \end{cases}$$

Find
(a) the values of the constants α and β,

(b) the probability that $1.5 \leqslant X \leqslant 2.5$,

(c) the probability density function $f(x)$.

Sketch the graph of $f(x)$, and hence, or otherwise, deduce the mean of X. Determine also the variance of X.

CAMB

Revision exercise D   Probability

Worked example

(a) In the *Who, What or Where* game, three contestants each choose one of the three categories of question. Assuming that the contestants choose independently and that they are equally likely to select any of the categories, find the probability that (i) all will choose the same category, (ii) all will choose different categories, (iii) two will be alike and the third different.

(b) From a pack of 52 cards, seven are taken at random, examined, and replaced. The cards are shuffled and another seven are drawn at random. Find the probability that at least one card will be drawn twice.

AEB

(a) Let A, B, and C represent each of the respective categories. Then ACB represents the event that the first contestant selects category A, the second category C and the third category B.

(i) P(all choose the same category)

$= P(AAA \text{ or } BBB \text{ or } CCC)$

$= P(AAA) + P(BBB) + P(CCC)$

$= \left(\frac{1}{3}\right)^3 + \left(\frac{1}{3}\right)^3 + \left(\frac{1}{3}\right)^3$

$= \frac{1}{9}.$

(ii) P (all will choose different categories)

$= P(ABC \text{ or } ACB \text{ or } BAC \text{ or } CAB \text{ or } CBA)$

$= 6 \times \left(\frac{1}{3}\right)^3$

$= \frac{2}{9}.$

(iii) The only other possibility is that two will be alike and the third different. So the required probability is $1 - \left(\frac{1}{9} + \frac{2}{9}\right) = 2/3.$

(b) We first find the probability that no card will be drawn twice. That is

$$^{45}C_7/^{52}C_7$$

$$= \frac{45!}{38!} \frac{45!}{7!} \frac{7!}{52!}$$

$$= \frac{45 \times 44 \times 43 \times 42 \times 41 \times 40 \times 39}{52 \times 51 \times 50 \times 49 \times 48 \times 47 \times 46}$$

$$= 0.339$$

Thus the required probability is

$$1 - 0.339 = 0.661.$$

Exercises

1. (a) A committee of four is to be selected from 10 candidates, 6 men (including Mr Brown) and 4 women (including Mrs Smith). In how many distinct ways can the committee be chosen?

    Assuming that each candidate is equally likely to be selected, determine the probabilities that the chosen committee contains

    (i) no women;
    (ii) two men and two women;
    (iii) Mr Brown, another man and any two women;
    (iv) Mrs Smith and any three others, excluding Mr Brown.

    (b) Three golfers A, B, C take turns at sinking a 2 m putt with A starting. If A misses, the ball is replaced and B tries and so on. If all three miss, A starts the cycle again. The winner is the first to sink the putt. Assuming that the probabilities that A, B, C sink any one putt are $1/3, 2/3, 2/3$ respectively, find for each player the probability that he wins. (You may quote the formula for the sum of an infinite geometric series.)

MEI

2. An experiment consists of tossing two unbiased dice one after the other. Let A be the event that the second die shows a 'six' and C be the event that two sixes are obtained. Calculate the following probabilities

$$P(A), P(B), P(C), P(A \cap B), P(A \cap C), P(C|A)$$

3. A prison governor gives condemned prisoners a choice of one of three chances to be freed. The prisoner has a fair coin which must be tossed normally, and he shall

either A: throw his second Head on his fourth toss,

or B: throw his third Head on his eighth or subsequent toss, or   *C:*   toss the coin four times, and record the number of Heads; toss it four more times, and get the same number of Heads as before.

For each of the three possible choices, find the probability that the prisoner is freed.

The governor observes that prisoners choose *A* and *B* equally frequently, and *C* twice as often as either of these. What proportion of prisoners is freed? Show that the probability that a prisoner has chosen *A*, given that he is freed, is 24/123.

OXFORD

4. A bag contains five different pairs of gloves. Two persons, *A* and *B*, take turns to draw a glove from the bag (without replacement), *A* drawing first. Find the probabilities
   (i) that the first glove drawn by *A* and the first glove drawn by *B* do not form a pair,
   (ii) that *A* obtains a pair in his first two draws,
   (iii) that *B* obtains a pair in his first two draws,
   (iv) that both *A* and *B* obtain a pair in their first two draws,
   (v) that at least one of the persons obtains a pair in his first two draws,
   (vi) that neither *A* nor *B* obtains a pair in his first two draws,
   (vii) that *B* completes a pair on making his second draw and *A* on making his third draw.

JMB

5. My post-man delivers mail on Monday at time *T* which is approximately Normally distributed with mean 8.50 am and with a variance of 225 minutes, and the number of letters *Y* is a Poisson variate with mean 2. Estimate the probability that
   (a) I receive just one letter on Monday,
   (b) the post-man arrives after I have left for work at 9 am,
   (c) that the post-man arrives between 7.40 am and 8.55 am with 3 or more letters,
   (d) given that he delivers more than three letters he actually delivers four letters.

6. If *A* and *B* are two events we know that

$$P(A \text{ or } B) = P(A) + P(B) - P(A \text{ and } B)$$

Find the corresponding results for the three events *A*, *B* and *C*.

7. The reliability of a system is the probability that it will function, within specified limits. Thus in the system below suppose that

the reliability of the component *A* is 0.8 and that of *B* is 0.9, then the reliability of the whole system is 0.8 × 0.9 providing the components are independent.

Find the reliability of the following systems where the reliability of the components are given in brackets.

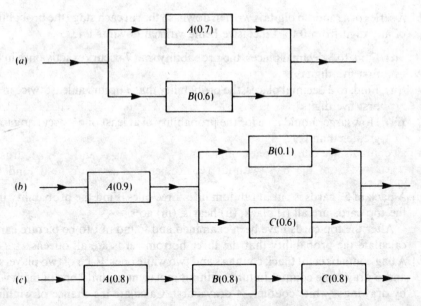

Owing to other considerations it was essential to increase the chain of components in (c) to six, each with the same reliability. What must their reliabilities be if the whole system is to be as reliable as the original.

8. Two friends agree to arrive at the village hall sometime between 7 pm and 8 pm and wait no longer than 15 minutes. If they arrive at any time within this one hour period find the probability that they will meet.

9. A coin is tossed $2n$ times. Find in terms of n the probability of obtaining an equal number of heads as tails.

Use Stirling's approximation

$$n! \simeq \sqrt{2\pi n} \cdot n^n \cdot e^{-n}$$

to show that for large n this probability is approximately $1/\sqrt{\pi n}$.

10. A and B are two mutually exclusive events such that $P(A) = 0.2$ and $P(B) = 0.8$. If, further, $P(C|A) = 0.3$ and $P(C|B) = 0.4$ find $P(A|C)$.

11. (a) Two unbiased dice are thrown. Find the probabilities of each of the following events:
    (i) the sum of the numbers showing is odd;
    (ii) there is at least one 6 showing;
    (iii) the sum of the numbers showing is odd *and* there is at least one 6 showing;
    (iv) the sum of the numbers showing is odd *or* there is at least one 6 showing (where *or* includes *both*);
    (v) the sum of the numbers showing is odd, given that there is at least one 6 showing.

(b) A series of n random digits is written down, so that at each stage the probability of any digit from 0 to 9 inclusive being written down is 0.1.

    (i) Find, to 3 decimal places, the probability that 7 occurs exactly once in the first five digits.

    (ii) Find, to 3 decimal places, the probability that 7 occurs at least twice in the first five digits.

    (iii) How large should n be for the probability of at least one 7 occurring to be greater than 0.9?

MEI

12. (a) A pack of 52 cards is cut at random into three piles. Find the probability that the top cards are all (i) black, (ii) hearts, (iii) aces.

    After the top cards have been examined and found not to be picture cards, calculate the probability that the three bottom cards are all queens.

  (b) A bag contains eight black counters and two white ones. Each of two players, A and B draws one counter in turn, without replacement, until one of them wins by drawing a white counter. A draws first. Calculate his chance of winning.

AEB

13. A particular species of bird lays from one to five eggs. The probability of r eggs is proportional to r. A collector of eggs takes one egg from a nest if there are more than three eggs there and none otherwise. Show that the probability that he takes an egg is 3/5.

In an area he finds six occupied nests of this species which have not previously been interfered with. Find, in fractional form, the probability that he obtains a total of m eggs from the six nests for $m = 0, 1, 2, 3, 4, 5$, and 6, respectively.

Obtain the mean of this distribution and determine the probabilities that he gets (i) fewer than four eggs, (ii) more than four eggs.

A second collector, obeying the same rule for collecting eggs, follows the first collector round the same six nests. Find the mean number of eggs the second collector would expect to collect.

JMB

14. An examination consists of two successive parts, of which the first must be passed before the second is attempted. Candidates may sit each part twice, and it may be assumed that any candidate who fails at his first sitting takes the opportunity of a resit. If the probability of passing the first part at any given sitting is p_1 and the probability of passing the second part at any given sitting is p_2, and all sitting are independent, write down the probability distribution of the total number of attempts and show that its mean is $(2 - p_1)[1 + p_1(2 - p_2)]$. (You should include candidates who ultimately fail as well as those who ultimately pass.) If $p_1 = p_2 = p$, find the value of p for which the expected number of attempts is maximised and show that the maximum expected number of attempts is $(34 + 14\sqrt{7})/27$.

OXFORD

15. Three men, A, B and C, share an office with a single telephone. Calls come in at random in the proportions 2/5 for A, 2/5 for B, 1/5 for C. Their work requires the men to leave the office at random times, so that A is out for half his working time and B and C each for a quarter of theirs.
For calls arriving in working hours, find the probabilities that
(a) no one is in to answer the telephone,
(b) a call can be answered by the person being called,
(c) three successive calls are for the same man,
(d) three successive calls are for three different men,
(e) a caller who wants B has to try more than three times to get him.

JMB

16. In Camelot it never rains on Friday, Saturday, Sunday or Monday. The probability that it rains on a given Tuesday is 1/5. For each of the remaining two days, Wednesday and Thursday, the conditional probability that it rains, given that it rained the previous day, is α, and the conditional probability that it rains, given that it did not rain the previous day, is β.
(a) Show that the (unconditional) probability of rain on a given Wednesday is $\frac{1}{5}(\alpha + 4\beta)$, and find the probability of rain on a given Thursday.
(b) If X is the event that, in a randomly chosen week, it rains on Thursday, Y is the event that it rains on Tuesday, and $\bar{Y}$ is the event that it does not rain on Tuesday, show that

$$P(X|Y) - P(X|\bar{Y}) = (\alpha - \beta)^2.$$

(c) Explain the implications of the case $\alpha = \beta$.

CAMB

17. In n independent trials with constant probability p of success at each trial, show that the ratio of the probabilities of $r + 1$ and r successes is

$$\frac{(n - r)p}{(r + 1)(1 - p)} \quad (0 \leqslant r \leqslant n - 1)$$

A manufacturer produces bracelets consisting of six sections linked together. After this linking up process each section independently has a probability q of having been scratched. The bracelets are inspected and those which are unscratched are sold for £2 each. Those with a single scratched section are sold as substandard for £1 each. Any with more than one scratched section are scrapped. It is found that the ratio of the number of bracelets sold as perfect to those sold as substandard is 4:1. Calculate q and hence find to three decimal places the proportion of the bracelets that are scrapped. Each bracelet costs £1 to produce. Find to the nearest £1 the expected profit on a batch which consists of 1000 bracelet before inspection.

JMB

18. A bird has flown into a room through an open window. The room has other windows of the same size and these are closed. The open window is the bird's only way out. It flies at windows until it finds the open one and escapes. Assuming that

the bird has no memory, so that each fresh attempt may be at any window, obtain the probability distribution of the number of attempts it makes in escaping.

The householder claims that his pet budgerigar would have had the intelligence not to try any window more than once. Assuming this to be true, obtain the corresponding probability distribution for the pet.

Find the probabilities that (a) the first bird would escape with fewer attempts than the pet, (b) the pet would take fewer attempts than the first bird.

JMB

19. (a) Every year very small numbers of American wading birds lose their way on migration between North and South America and arrive in Great Britain instead, so that in September the proportion of American waders amongst the waders in Great Britain is about one in ten thousand.

   At Dunsmere (a bird reserve in Great Britain), one September, there are twenty thousand waders, which may be regarded as a random sample of the waders present in Great Britain. Determine the probability that there are
   (i) no American waders present at Dunsmere,
   (ii) more than two American waders present at Dunsmere.

   (b) Three quarters of all the sightings in Great Britain of American waders are made in the autumn. Suppose that in 1980 there will be ten sightings of American waders at Dunsmere. Assuming that all sightings are independent of one another, determine the probability that exactly seven of these sightings will be made in the autumn.

CAMB

20. At a horse show there is a jumping competition in which each horse attempts to clear 4 successive fences, each fence being more difficult than the previous one. A particular horse has probability $(5 - i)/5$ of clearing the ith fence ($i = 1, 2, 3, 4$), independent of its performance over the other fences.
   (a) Find the probability that the horse clears all four fences.
   (b) Find the probability that the horse does not clear any fence.
   (c) Determine the expected value of the number of fences that the horse clears.

CAMB

Appendix

Table 1 The normal distribution

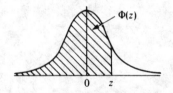

(a) *Distribution Function.* The tabulated value is
$\Phi(z) = P(Z < z)$ where Z is the standardized
Normal random variable, $N(0, 1)$.

z	.00	.01	.02	.03	.04	.05	.06	.07	.08	.09
.0	.5000	.5040	.5080	.5120	.5160	.5199	.5239	.5279	.5319	.5359
.1	.5398	.5438	.5478	.5517	.5557	.5596	.5636	.5675	.5714	.5753
.2	.5793	.5832	.5871	.5910	.5948	.5987	.6026	.6064	.6103	.6141
.3	.6179	.6217	.6255	.6293	.6331	.6368	.6406	.6443	.6480	.6517
.4	.6554	.6591	.6628	.6664	.6700	.6736	.6772	.6808	.6844	.6879
.5	.6915	.6950	.6985	.7019	.7054	.7088	.7123	.7157	.7190	.7224
.6	.7257	.7291	.7324	.7357	.7389	.7422	.7454	.7486	.7517	.7549
.7	.7580	.7611	.7642	.7673	.7704	.7734	.7764	.7794	.7823	.7852
.8	.7881	.7910	.7939	.7967	.7995	.8023	.8051	.8078	.8106	.8133
.9	.8159	.8186	.8212	.8238	.8264	.8289	.8315	.8340	.8365	.8389
1.0	.8413	.8438	.8461	.8485	.8508	.8531	.8554	.8577	.8599	.8621
1.1	.8643	.8665	.8686	.8708	.8729	.8749	.8770	.8790	.8810	.8830
1.2	.8849	.8869	.8888	.8907	.8925	.8944	.8962	.8980	.8997	.9015
1.3	.9032	.9049	.9066	.9082	.9099	.9115	.9131	.9147	.9162	.9177
1.4	.9192	.9207	.9222	.9236	.9251	.9265	.9279	.9292	.9306	.9319
1.5	.9332	.9345	.9357	.9370	.9382	.9394	.9406	.9418	.9429	.9441
1.6	.9452	.9463	.9474	.9484	.9495	.9505	.9515	.9525	.9535	.9545
1.7	.9554	.9564	.9573	.9582	.9591	.9599	.9608	.9616	.9625	.9633
1.8	.9641	.9649	.9656	.9664	.9671	.9678	.9686	.9693	.9699	.9706
1.9	.9713	.9719	.9726	.9732	.9738	.9744	.9750	.9756	.9761	.9767
2.0	.9772	.9778	.9783	.9788	.9793	.9798	.9803	.9808	.9812	.9817
2.1	.9821	.9826	.9830	.9834	.9838	.9842	.9846	.9850	.9854	.9857
2.2	.9861	.9864	.9868	.9871	.9875	.9878	.9881	.9884	.9887	.9890
2.3	.9893	.9896	.9898	.9901	.9904	.9906	.9909	.9911	.9913	.9916
2.4	.9918	.9920	.9922	.9925	.9927	.9929	.9931	.9932	.9934	.9936
2.5	.9938	.9940	.9941	.9943	.9945	.9946	.9948	.9949	.9951	.9952
2.6	.9953	.9955	.9956	.9957	.9959	.9960	.9961	.9962	.9963	.9964
2.7	.9965	.9966	.9967	.9968	.9969	.9970	.9971	.9972	.9973	.9974
2.8	.9974	.9975	.9976	.9977	.9977	.9978	.9979	.9979	.9980	.9981
2.9	.9981	.9982	.9982	.9983	.9984	.9984	.9985	.9985	.9986	.9986
3.0	.9987	.9987	.9987	.9988	.9988	.9989	.9989	.9989	.9990	.9990
3.1	.9990	.9991	.9991	.9991	.9992	.9992	.9992	.9992	.9993	.9993
3.2	.9993	.9993	.9994	.9994	.9994	.9994	.9994	.9995	.9995	.9995
3.3	.9995	.9995	.9995	.9996	.9996	.9996	.9996	.9996	.9996	.9997
3.4	.9997	.9997	.9997	.9997	.9997	.9997	.9997	.9997	.9997	.9998

(b) *Upper Percentage Points.* The tabulated value of z_p, where $P(Z > z_p) = p$, so that $1 - \Phi(z_p) = p$.

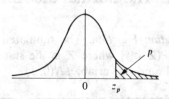

p	0.05	0.025	0.01	0.005	0.001	0.0005
z_p	1.64	1.96	2.33	2.58	3.09	3.29

Table 2   Random digits

The tabulated digits are independent observations from a distribution in which the digits 0–9 have equal probabilities.

90159	96465	96858	73457	74257	19970	87916	89204	08631	82959
63835	94333	80384	57737	41936	82487	55651	58651	72611	03774
83793	02044	76074	78034	19256	29922	56192	86943	19716	27732
37954	93470	67333	52645	49638	02802	74126	57816	68432	06050
52203	80860	02071	28397	13988	16152	63299	62855	93960	03357
69684	84874	19256	74311	14029	66457	64647	98847	01753	43677
41089	71096	43777	31817	58684	36247	04775	88407	80496	32094
03015	41571	08057	85319	09346	17075	03058	91232	23473	95982
77980	81547	74092	32879	60549	30487	02476	43878	80824	90470
50512	38559	17847	31353	01305	67204	47248	65847	39103	78374
49195	72090	22959	61455	14242	08028	09064	88436	19695	92076
81070	06070	02688	30084	03248	56913	46961	11143	67229	96523
00237	39550	75537	04273	43291	69091	03682	14784	50468	67799
43764	21488	62091	20761	69330	07661	61564	75202	01854	36385
10350	16762	67504	11431	87820	99979	84539	98135	83516	02353
25163	41056	64314	87456	55548	62706	03370	01131	84842	68637
17533	12864	00959	95443	54257	97194	30811	98350	44868	76917
16187	08144	03848	65933	88249	90334	31860	09413	11770	63679
66049	15483	80910	09046	83696	73702	32145	30374	42841	47797
25002	07562	11250	27858	72007	85400	55906	12765	31490	91355

Table 3 Values of the χ^2 distribution exceeded with probability P

P d.f.	0.995	0.975	0.050	0.025	0.010	0.005	0.001
1	3.9×10^{-5}	9.8×10^{-4}	3.84	5.02	6.63	7.88	10.83
2	0.010	0.051	5.99	7.38	9.21	10.60	13.81
3	0.071	0.22	7.81	9.35	11.34	12.84	16.27
4	0.21	0.48	9.49	11.14	13.28	14.86	18.47
5	0.41	0.83	11.07	12.83	15.09	16.75	20.52
6	0.68	1.24	12.59	14.45	16.81	18.55	22.46
7	0.99	1.69	14.07	16.01	18.48	20.28	24.32
8	1.34	2.18	15.51	17.53	20.09	21.96	26.13
9	1.73	2.70	16.92	19.02	21.67	23.59	27.88
10	2.16	3.25	18.31	20.48	23.21	25.19	29.59
11	2.60	3.82	19.68	21.92	24.73	26.76	31.26
12	3.07	4.40	21.03	23.34	26.22	28.30	32.91
13	3.57	5.01	22.36	24.74	27.69	29.82	34.53
14	4.07	5.63	23.68	26.12	29.14	31.32	36.12
15	4.60	6.26	25.00	27.49	30.58	32.80	37.70
16	5.14	6.91	26.30	28.85	32.00	34.27	39.25
17	5.70	7.56	27.59	30.19	33.41	35.72	40.79
18	6.26	8.23	28.87	31.53	34.81	37.16	42.31
19	6.84	8.91	30.14	32.85	36.19	38.58	43.82
20	7.43	9.59	31.41	34.17	37.57	40.00	45.32
21	8.03	10.28	32.67	35.48	38.93	41.40	46.80
22	8.64	10.98	33.92	36.78	40.29	42.80	48.27
23	9.26	11.69	35.17	38.08	41.64	44.18	49.73
24	9.89	12.40	36.42	39.36	42.98	45.56	51.18
25	10.52	13.12	37.65	40.65	44.31	46.93	52.62
26	11.16	13.84	38.89	41.92	45.64	48.29	54.05
27	11.81	14.57	40.11	43.19	46.96	49.64	55.48
28	12.46	15.31	41.34	44.46	48.28	50.99	56.89
29	13.12	16.05	42.56	45.72	49.59	52.34	58.30
30	13.79	16.79	43.77	46.98	50.89	53.67	59.70
40	20.71	24.43	55.76	59.34	63.69	66.77	73.40
50	27.99	32.36	67.50	71.42	76.16	79.49	86.66
60	35.53	40.48	79.08	83.30	88.38	91.95	99.61
70	43.28	48.76	90.53	95.02	100.43	104.22	112.32
80	51.17	57.15	101.88	106.63	112.33	116.32	124.84
90	59.20	65.65	113.15	118.14	124.12	128.30	137.21
100	67.33	74.22	124.34	129.56	135.81	140.17	149.44

For degrees of freedom $f > 100$, test $\sqrt{2\chi^2_{(f)}}$ as $N(\sqrt{2f-1}, 1)$.

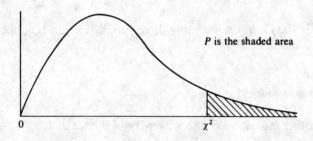

P is the shaded area

Answers to exercises

Exercise 1

1.

0	1	2	3	4	5	6
15	12	11	8	7	5	2

2.

0	1	2	3	4	5	6	7
1	4	5	12	10	5	10	3

3.   (i) 10.45 cm, 11.45 cm, 12.45 cm, 13.55 cm,
    (ii) 10.95 cm, 11.95 cm, 13 cm,
    (iii) 1 cm, 1 cm, 1.1 cm.

4. 1, 1, 3, 10, 7, 13, 2, 9, 3, 1.

8. It depends on the use to which the component is put. However B has fewer early failures.

9. 4.45 kg.

17. (*b*) 283, 184, 169, 14.

18. An average 15 year-old girl in 1970 could expect to live a further 61.5 years.

Exercise 2.1

1. 4; 0, 5; 4.5.

2. 62, 64, 64.

3. 6.

4. (i) When data arranged symmetrically about a particular value – but consider 1, 2, 3, 3, 3, 6.

5. Yes; mean is 100.17, but notice only three people have an IQ above 100!

6. 44.4, 3.

7. Median: to save money!

8. Mean – another indication of the sensitivity of the mean.

9. 4.4; (i) 14.4, (ii) −0.6, (iii) $a + 4.4$, (iv) 44, (v) $4.4 \times b$, (vi) $a + 4.4 \times b$.

10. 3.43, 2, 3.

11. 9.23.

12. £6700.

Exercise 2.2

1. 9.4.

2. (a) 31.3,   (b) 158.03.

3. (i) 4.26,   (ii) 4.26.

4. 40.28.

6.

10	20	30	40	50	60	70	80	90
16	13	6	7	4	7	0	0	7

   At first glance it appears to be the 50p type (among those sold at all). However he takes rather less money from the 10p and 30p types!

7. 3 in each case. One can differentiate on the mean suggesting type B offers the best average yield.

8. 1.442.

9. (a) 5.516,   (b) 8275,   (c) 16.4128.

Exercise 2.3

1. 24.4 mph, 20 mph; 10% assuming a 30 mph speed limit.

2. 27.0,   25.

3. 5.10.

4. 0.45 mins.

5. 0.309.

6. 16.736.

Exercise 2.4

1. (a) £107   (b) 14%.

2. (a) 31   (b) 36.

3. 52;   32.5–61.5.

4. 69.5 g,   74.5 g,   65 g.

5. 34 yrs; $\simeq 18\%$.

Exercise 3.1

1. 0.86.

2. (i) 1,   (ii) 0,   (iii) $\frac{1}{2}$,   (iv) $\frac{1}{2}$.

3. (i) 0.37,   (ii) 0.28,   (iii) 0.88.

5. (i) $\frac{1}{2}$,   (ii) $\frac{1}{3}$,   (iii) 1,   (iv) 0.

6. (i) $\frac{1}{8}$,   (ii) $\frac{1}{150}$.

7. (i) $\frac{1}{2}$,   (ii) $\frac{1}{4}$,   (iii) $\frac{1}{10}$,   (iv) $\frac{1}{52}$,   (v) $\frac{12}{13}$,   (vi) $\frac{1}{26}$,   (vii) $\frac{2}{13}$,   (viii) $\frac{11}{13}$.

8. (i) $\frac{1}{4}$ (ii) even.

9. (i) $\frac{1}{5}$ (ii) $\frac{2}{5}$.

Exercise 3.2

1. (a) $\frac{2}{9}$,   (b) $\frac{1}{6}$,   (c) $\frac{1}{9}$,   (d) $\frac{1}{36}$,   (e) $\frac{5}{36}$.

2. (a) $\frac{1}{2}$,   (b) $\frac{4}{13}$,   (c) $\frac{2}{13}$,   (d) $\frac{1}{26}$.

3. $\frac{1}{12}$.

4. (i) (a) $\frac{1}{2}$,   (b) $\frac{3}{10}$,   (c) $\frac{1}{100}$,   (d) $\frac{1}{10}$;

   (ii) (a) $\frac{1}{2}$,   (b) $\frac{3}{10}$,   (c) $\frac{1}{90}$,   (d) 0.

5. (a) $\frac{1}{36}$,   (b) $\frac{1}{18}$,   (c) $\frac{5}{9}$.

6. (a) $\dfrac{1}{16}$, (b) $\dfrac{1}{8}$, (c) $\dfrac{7}{16}$.

8. (a) $\dfrac{3}{8}$, (b) $\dfrac{1}{8}$, (c) $\dfrac{7}{16}$, (d) $\dfrac{5}{64}$.

9. (a) $\dfrac{1}{5}$, (b) $\dfrac{1}{5}$, (c) $\dfrac{2}{5}$, (d) $\dfrac{2}{5}$, (e) $\dfrac{2}{5}$, (f) $\dfrac{3}{10}$.

10. (a) $\dfrac{4}{7}$, (b) $\dfrac{1}{7}$, (c) $\dfrac{6}{49}$, (d) $\dfrac{24}{49}$.

11. (a) 216, (b) 6, (c) $\dfrac{1}{36}$.

12. $52 \times 51 \times 50 \times 49 = 6\,497\,400$, $\dfrac{1}{6\,497\,400}$.

Exercise 3.3

1. (a) $\dfrac{6!}{3!}$, (b) $\dfrac{n!}{(n-2)!}$, (c) $\dfrac{9!}{6!\,3!}$, (d) $\dfrac{n!}{(n-4)!}$, (e) $\dfrac{n!}{(n-r)!}$, (f) $\dfrac{n!}{(n-r)!\,r!}$

2. (a) 720, (b) 42, (c) 5814, (d) 56.

3. (a) $n(n+1)$, (b) $(n+1)(n+2)$, (c) $n-1$.

4. (a) $\dfrac{1}{256}$, (b) $\dfrac{1}{32}$.

5. 0.0001.

6. (a) $\dfrac{1}{336}$, (b) $\dfrac{1}{56}$.

7. (a) $\dfrac{1}{35}$, (b) $\dfrac{1}{35}$.

8. (a) $\dfrac{1}{12^4} = \dfrac{1}{20736}$, (b) $\dfrac{1}{12^3} = \dfrac{1}{1728}$, (c) $\dfrac{1}{12 \times 11 \times 10 \times 9} = \dfrac{1}{11\,880}$.

9. No: number of days per month varies, besides other factors involved.

10. (a) $\dfrac{1}{6!} = \dfrac{1}{720}$, (b) $\dfrac{1}{12}$.

Exercise 3.4

1. (a) 60, (b) 15, (c) 1, (d) 1, (e) 7, (f) 28,
   (g) 28, (h) 969, (i) n, (j) $n!$, (k) $n(n-1)$, (l) 12.

2. (a) 840,   (b) 35.

4. (a) $^{n+m}C_n = {^{n+m}C_m}$,   (b) $^{n+1}C_{r-1} = {^{n+1}C_{n-r+2}}$.

5. 50 400.

6. 792, $\dfrac{99}{512}$.

7. 1960, $\dfrac{3}{14}$.

8. 6.

9. $^{48}C_9/^{52}C_{13} \simeq 0.003$.

11. (a) 120,   (b) 216.

12. (a) $\dfrac{43}{91}$,   (b) 1.

Exercise 3.5

1. (a) $\dfrac{1}{7}$,   (b) $\dfrac{2}{7}$,   (c) $\dfrac{4}{7}$,   (d) $\dfrac{3}{7}$.

2. (a) $\dfrac{1}{20}$,   (b) $\dfrac{1}{10}$,   (c) $\dfrac{1}{4}$,   (d) $\dfrac{23}{50}$.

3. (a) $\dfrac{64}{52 \times 51 \times 50} \simeq 0.00048$,   (b) $\dfrac{96}{52 \times 51} \simeq 0.036$.

4. (a) $\dfrac{6}{11}$,   (b) $\dfrac{1}{22}$,   (c) $\dfrac{9}{22}$.

5. (a) $\dfrac{5}{33}$,   (b) $\dfrac{5}{44}$,   (c) $\dfrac{1}{220}$.

6. (i) $\dfrac{1}{11}$;   (ii) (a) $\dfrac{3}{11}$,   (b) $\dfrac{4}{11}$.

7. (i) (a) $\dfrac{4}{9}$,   (b) $\dfrac{2}{9}$;   (ii) (a) $\dfrac{4}{9}$,   (b) $\dfrac{1}{6}$.

8. (a) $\dfrac{27}{512}$,   (b) $\dfrac{1}{512}$,   (c) $\dfrac{17}{64}$.

9. (a) $\dfrac{1}{22}$,   (b) $\dfrac{2}{11}$.

10. (a) $\dfrac{671}{1296}$,   (b) $1 - \left(\dfrac{5}{6}\right)^n$.

11. (a) $\dfrac{m}{m+n}$, (b) $\dfrac{mn}{(m+n)(m+n-1)}$, (c) $\dfrac{2mn}{(m+n)(m+n-1)}$, (d) $\dfrac{m}{m+n-1}$.

Exercise 4.1

2. (a) 6, $3\frac{1}{3}$, $18\frac{1}{3}$, 4.28;
   (b) 11.2, 4.2, 27.56, 5.25;
   (c) 5.5, 2.45, 8.25, 2.9.

3. (a) 4.25.

4. 12.5.

5. Note that $(A - \bar{x})^2$ is never negative.

7. (a) 3.12, (b) 1.25.

10. (a) 1, (b) 0.89, (c) 0.89.

Exercise 4.2

1. (a) 2.6, (b) 4.5, (c) 11.1, 0.42.

2. 4.5, 2.87.

5. 0.525, 0.55.

6. 7.02, 2.37.

7. (a) 31.8, 10.85 (b) 51.8, 10.85 (c) 38.16, 13.02.

8. (a) 7, 1.29, (b) 6.09, 1.93.

10. $\left(\dfrac{N_1 S_1^2 + N_2 S_2^2}{N_1 + N_2} + \dfrac{N_1 N_2 (M_1 - M_2)^2}{(N_1 + N_2)^2} \right)^{1/2}$.

11. 12.08 yrs, 1.83 yrs.

12. 1.32 kg, 1.07 kg^2, 1.03 kg, 1.3 kg.

Exercise 4.3

1. 101–116, 24.6–52.6, 32.5–61.5, 65–73.25, 15.76–55.1.

2. (a) 17.1 kg, 4.1 kg, (b) 17.85 kg, 5.57 kg;
   median and semi-interquartile range.

3. $52°$, $7.7°$, $53.2°$, $10.66°$.

4. 64.09 yrs, 66.5 yrs, 74.68 yrs, 55.73 yrs.

Exercise 5.1

1. F_1 and F_2 are independent and G_1 and G_2 are mutually exclusive.

2. (i) 0,   (ii) $\frac{1}{8}$,   (iii) $\frac{1}{8}$,   (iv) 0,   (v) $\frac{1}{4}$,   (vi) $\frac{1}{2}$;

    (a) A and B,   (b) A and C.

3. 0.42, 0.46, 0.12, 0.

4. (a) $\frac{1}{125}$,   (b) $\frac{1}{5}$,   (c) $\frac{61}{125}$.

5. (a) (i) $\frac{2}{3}$,   (ii) $\frac{1}{60}$,   (iii) $\frac{1}{12}$;

    (b) (i) $\frac{1}{9}$,   (ii) $\frac{121}{144}$.

6. (a) 0.064,   (b) 0.72,   (c) 0.01.

7. $\frac{11}{12}$;   no!

8. (a) $p_1 p_2$,   (b) $p_1(1 - p_2)$,   (c) $p_1 + p_2 - 2p_1 p_2$,   (d) $1 - p_1 - p_2 + p_1 p_2$.

9. (a) 0.2,   (b) 0.33,   (c) 0.06,   (d) 0.6.

10. (a) $\frac{1}{3}$,   (b) $\frac{2}{9}$,   (c) $\frac{1}{3}$,   (d) $\frac{1}{3}$,   (e) $\frac{2}{3}$.

12. $\left(\frac{1}{8}\right)^5 \simeq 0.00003$.

13. (a) $\frac{57}{110}$,   (b) $\frac{4}{55}$.

14. (a) $\left(\frac{1}{2}\right)^{10}$,   (b) $10\left(\frac{1}{2}\right)^{10}$.

15. (a) 1,   (b) $\frac{1}{2}$,   (c) $\frac{1}{13}$.

16. (a) If one of the events cannot occur.
    (b) Only if at least one of the events cannot occur.
    (c) Whenever $P(A) \neq P(B)$.

17. $\dfrac{7}{9}$.

18. $\dfrac{1}{2}$.

19. (a) $\dfrac{10}{21}$, (b) $\dfrac{2}{7}$.

20. (a) $\dfrac{25}{36}$, (b) $\dfrac{2}{3}$.

21. (a) $\dfrac{5}{8}$, (b) $\dfrac{7}{8}$, (c) $\dfrac{3}{8}$, (d) $\dfrac{3}{8}$, (e) $\dfrac{1}{2}$.

22. (a) $\dfrac{1}{12}$, (b) $\dfrac{1}{4}$, (c) $\dfrac{1}{3}$, (d) $\dfrac{3}{4}$, (e) $\dfrac{1}{4}$.

23. (a) $\dfrac{5}{12}$, (b) $\dfrac{8}{15}$, (c) $\dfrac{1}{12}$.

24. (a) (i) $\dfrac{4}{17}$, (ii) $\dfrac{1}{17}$, (iii) $\dfrac{5}{17}$;

(b) (i) $\dfrac{1}{4}$, (ii) $\dfrac{1}{13}$, (iii) $\dfrac{4}{13}$;

(c) $\dfrac{17}{108}$.

25. (a) 0.4; (b) 0.615; (c) (i) 0.033, (ii) 0.075, (iii) 0.444.

26. (i) $\dfrac{2}{261}$, (ii) 0.026; 0.708, 0.201

Exercise 6.1

1. (a) 2.39, (b) 0.2, (c) $3\frac{1}{3}$; 4.2, -0.6, $5\frac{2}{3}$;

   (i) 7.16, 0.6, 10; (ii) 7.55, -1.2, $11\frac{1}{3}$; (iii) 7.88, 2, $14\frac{1}{3}$

2. (a) 3.5, 19.3, 7.05; (b) 4.5, 63.45.

3. (a) 3p (b) 30p, 3p.

4. $33\frac{1}{3}$p; independence.

5. $3\frac{3}{8}$, $2\frac{31}{64} \simeq 2.48$.

6. (a) 2, 3; (b) 4, 4.

7. $- £\frac{1}{6} \simeq 16.5\text{p}$ lost.

Exercise 6.2

1. (a) 3μ, (b) $2\mu + 1$, (c) $\sigma^2 + \mu^2$, (d) $3\sigma^2 + 3\mu^2 + 5\mu + 9$, (e) $25\sigma^2$, (f) $9\sigma^2$, (g) $16\sigma^2$
(h) $\sigma^2 + \mu^2 + 2\mu$.

3. (a) 10μ,   (b) $4\mu - 6$,   (c) $16\sigma^2$,   (d) 0,
   (e) $9\sigma^2$,   (f) σ^2,   (g) σ^2,   (h) $4\sigma^2$.

5. $\frac{32}{31}$; $\frac{24}{31}$, $1\frac{26}{31}$.

6. $1\frac{7}{8}$.

7. 4.71,   6.30.

Exercise 7.1

2. (a) 0.233,   (b) 0.372,   (c) 0.395.

3. 0.375;   assuming $P(\text{Boy}) = P(\text{Girl}) = \frac{1}{2}$.

4. (a) 0.329,   (b) 0.018,   (c) 0.302.

5. (a) 0.531,   (b) 0.469.

6. (i) $\frac{4}{9}$,   (b) $\frac{1}{9}$ (ii) 0.228.

7. (i) 0.107,   (ii) 0.376,   (iii) 0.302,   (iv) 0.322.

8. (a) 0.000475,   (b) 0.265.

9. (a) (i) 0.132,   (ii) 0.132,   (b) (i) 0.0714,   (ii) 0.0714.

10. 5.

11. 30, $\frac{1}{3}$.

12. At least 13 times.

13. (a) 0.00098,   (b) 0.246,   (c) 0.205.

15. (a) 3.2,   (b) 0.598.

16. 1.65,   1.05.

19. 0.579,   0.347,   0.069,   0.005;   a loss of approximately 2p.

20. 16.5,   42.4,   45.4,   25.9,   8.3,   1.4,   0.1.

Exercise 7.2

1. $\frac{1}{3}$.

3. $\frac{2}{3}$,   22p.

4. (a) $P(X = r) = \left(\frac{1}{6}\right)^{r-1} \frac{5}{6}$,   for $r = 1, 2, 3, \ldots$;

   (i) $\frac{1}{3}$,   (ii) $\frac{1}{216}$,   (iii) 0;

   (b) (i) $\frac{25}{36}$,   (ii) 6.

5. (a) 0.684,   (b) 0.1406.

Revision exercise A

1. 35.1 yrs,   17.6 yrs.

2. 7.773 half-days,   73.40 half-days^2.

3. (i) 2, 0, 1, 0, 4, 4, 8, 5, 8, 14, 12, 5, 0, 0, 1;

   (ii) $\bar{n} \simeq 34.44$, $S \simeq 7.88$;   (iii) 74.3%.

4. (b) (i) $\frac{1}{4}$,   (ii) $\frac{25}{102}$.

5. The first.

6. (a) $\frac{12}{19}$,   (b) $\frac{3}{95}$.

7. (a) 0.1157,   0.1250;   (b) 0.5177,   0.4914;   (c) 0.6651,   0.6186.

8. (a) $\frac{1}{21}$,   (b) $\frac{2}{7}$,   (c) $\frac{4}{9}$,   (d) $\frac{5}{42}$,   (e) 1,   (f) 1.33,   0.

9. (a) $\frac{1}{3}$,   (b) $\frac{2}{9}$,   (c) $\frac{16}{243}$;   50p, $\frac{16}{81}$,   0.177.

10. (a) £2.25,   (b) £0.857.

11. (a) 12.25p,   (b) 0.818   (c) $\frac{47}{66} \simeq 0.712$.

12. 0.228,   0.209.

13. 6.49,   1.71,   7.

14. Type B.

Exercise 8.1

1. $\frac{2}{243}$.

2. $\frac{8}{3}$, $\frac{8}{9}$;   0.879.

3. $\frac{1}{2}$,   3, $\frac{1}{3}$.

4. 5,   5, $8\frac{1}{3}$.

5. (i) $\frac{a}{2}$, $\frac{a^2}{12}$;   (ii) $\frac{a + b}{2}$, $\frac{(b - a)^2}{12}$.

6. 2.

7. $\frac{4}{9}$.

8. $\frac{1}{144}$, $2\frac{1}{4}$, $\frac{17}{432} \simeq 0.039$.

9. (a) 0.741,   (b) 1.155,   (c) 1.265.

10. (i) 4,   (ii) 1.231.

11. 2.5,   0.185.

12. (b) $\frac{1}{4}$,   3.055,   (c) 0.528.

13. 1;   (i) 3.5,   (ii) 7.083,   (iii) 0.5.

14. (a) 1, $y_1 = 1 - x, y_2 = 1 + x$;
    (b) 0.408 mm,   (c) $\frac{1}{4}$.

15. 0.542,   4,   5.07,   5.33.

16. (a) 2, (b) $f(x) = 2$, $0 \leqslant x \leqslant 0.5$, (c) 0.25 m, (d) 0.144 m.

17. (i) $\dfrac{12}{5}$, $1\frac{1}{2}$; (ii) 0.125.

18. $\dfrac{1}{108}$, 2.4, 2, 1.44, 1.2.

19. $\mu = 2\frac{1}{2}$; 0.803; 0.456.

Exercise 9.1

1. (i) (a) 0.9032, (b) 0.0968, (c) 0.8064;
    (ii) (a) 0.9821, (b) 0.0179, (c) 0.9642;
    (iii) (a) 0.9673, (b) 0.0327, (c) 0.9346.

2. (a) 1.6, (b) -1.3, (c) -1.53.

3. (a) 0.9332, (b) 0.9938, (c) 0.5, (d) 1, (e) 0.3085, (f) 0.0228, (g) 0.3944, (h) 0.3944,
   (i) 0.6247, (j) 0.2761.

4. (a) 0.1198, (b) 0.8802, (c) 0.3778, (d) 0.2646, (e) 0.1061.

5. (a) 0, (b) 0.675, (c) 1.96, (d) -0.647, (e) 1.645.

6. (a) 5, (b) 6.35, (c) 7.073, (d) 6.684, (e) 1.71.

7. (a) 0.0228, (b) 0.8174, (c) 0.0038; 113.

8. (a) 95%, (b) 0.675.

9. (i) 0.1056, (ii) 0.3085, (iii) 0.5328.

10. 0.4602.

11. (a) 0.150, (b) 0.924, (c) 0.651.

12. 0.0098, 0.971, 6.19.

Exercise 9.2

1. 3.67 kg, 0.0037.

2. 0.78 kg.

3. 0.0052.

4. (a) 0.62%, (b) 4.292 g.

5. 43.78 cm,   7.79 cm^2.

6. 252.5 g.

7. 3.8%,   54.19 g.

8. (a) 73,   (b) 75.

9. 35 min,   3.162 min,   0.26,   39.05 min.

10. −0.675,   0.675,   53%,   371.

11. 383.

12. 8 cm,   1.157 cm,   6.10–9.90.

13. 137,   149.5.

Exercise 9.3

1. (a) 0.656,   (b) 0.658.

2. (a) 0.162,   (b) 0.039.

3. 0.0885.

4. 0.749.

5. (a) about 18,   (b) about 24.

6. Approximately 204.

7. 0.011.

8. (a) 0.128,   (b) 0.742.

9. (a) 0.9774,   (b) 0.00.

10. (a) 0.6247,   (b) 93.32%,   (c) 0.7852.

11. $\sum_{r=70}^{100} {}^{100}C_r \left(\frac{1}{4}\right)^{100-r} \left(\frac{3}{4}\right)^r$,   (b) 0.898.

12. $\frac{2000!}{N!(2000-N)!} \left(\frac{1}{30}\right)^N \left(\frac{29}{30}\right)^{2000-N}$;   86 lines;   2 × 47 lines > 86 lines.

13. $np(1-p) > 1$;   (a) almost certain,   (b) 0.980.

Exercise 10.1

1. (a) $3\frac{1}{2}$, (b) 2, (c) $2\frac{11}{12}$, (d) 1, (e) $5\frac{1}{2}$, (f) $3\frac{11}{12}$, (g) 5, (h) $12\frac{2}{3}$, (i) $15\frac{1}{6}$, (ii) 5.

2. (a) $1\frac{1}{4}$, (b) $16\frac{1}{2}$, (c) 125.

3. (a) $\mu + 3$, (b) $3\sigma^2$ (c) $3\sigma^2$ (d) $17\sigma^2$ (e) $18\sigma^2$, (f) $a^2\sigma^2 + 2b^2\sigma^2$.

4. 7, $17\frac{1}{2}$.

5. The first boy: it requires independence to add the variances.

6. p, pq.

7. $\sigma^2 + \mu^2 + 2a\mu + a^2$.

Exercise 10.2

1. (i) $\mu_1 + \mu_2$; (ii) $\sigma_1^2 + \sigma_2^2 + 2\sigma_1\sigma_2\rho$; approximately 12%.

2. (i) $(a^2 + b^2)\sigma^2$, (ii) $2ab\sigma^2$.

5. $\mathrm{Var}(X) - \mathrm{Var}(Y)$.

6. $\dfrac{t}{6}\left(1 - \dfrac{5}{6}t\right)^{-1}$; (a) 6, (b) 30.

7. $\dfrac{4}{7}$; $\dfrac{7t}{4}\left(1 - \dfrac{9t}{16}t\right)^{-1}$, $2\frac{2}{7}$, 1.57.

9. $(0.55 + 0.45t)^4$, 4, $4 \times 0.55 \times (0.45)^3$.

Exercise 11.1

1. 2.7182.

2. (a) $1 + 2 + 2 + 1\frac{1}{3}\ldots\dfrac{2^r}{r!}$,

   (b) $1 + 2x + 2x^2 + \dfrac{4}{3}x^3 \ldots \dfrac{(2x)^r}{r!}$,

   (c) $1 - x + \dfrac{x^2}{2} - \dfrac{x^3}{6} \ldots \dfrac{(-x)^r}{r!}$

   (d) $1 - a + \dfrac{a^2}{2} - \dfrac{a^3}{6} \ldots \dfrac{(-a)^r}{r!}$.

3. (a) 20.085, (b) 4.953, (c) 0.100, (d) 0.326.

Exercise 11.2

1. (a) e^{-a}, (b) $e^{-a}\dfrac{a^5}{5!}$, (c) $\dfrac{e^{-a}a^{r+1}}{(r+1)!}$.

2. (a) 0.018, (b) 0.433, (c) 0.762.

3. (a) 0.194, (b) 0.184.

4. 0.762.

5. 0.735.

6. (a) 0.189, (b) 0.766.

7. (a) 0.017, (b) 0.084, (c) 0.161.

8. 1.93,   2.005; 15, 28, 27, 17, 8, 3 to the nearest integer.

9. 108.7,   66.3,   20.2,   4.1,   0.6,   a random process.

10. (a) 0.050, (b) 0.185; 8.

11. 0.0141.

12. (i) 0.19 (ii) 0.21; 0.012.

13. (a) $P(Y = r) = \dfrac{e^{-a}a^{r-1}}{(r-1)!}$,   for $r = 1, 2, 3, \ldots$; $a + 1$, a;

   (b) $P(Z = r) = \dfrac{1}{(1 - e^{-a})}\left[\dfrac{e^{-a}a^r}{r!}\right]$,   for $r = 1, 2, 3, \ldots$;

   $\dfrac{a}{1 - e^{-a}}$, $\dfrac{a}{1 - e^{-a}} + \dfrac{a^2}{1 - e^{-a}} - \dfrac{a^2}{(1 - e^{-a})^2}$

Exercise 11.3

1. (a) 0.353, (b) 0.1576, (c) 0.0025, (d) 0.135.

2. 0.521, 0.50; 28.5, 14.9, 3.9, 0.7; 0.002.

3. 3.

5. 0.423.

6. (a) 0.018, (b) 0.084, (c) 0.000.

7. 0.301.

8. (i) 22 days, (ii) 19 days; particular lawnmower not used on 39 days.

9. (i) 0.191, (ii) 0.018, (iii) 0.264.

10. (i) 0.0228, (ii) 0.9772.

11. (i) 9, (ii) 26, (iii) 43; 8.

12. (i) 0.082, (ii) 0.242; 6.155 s.

13. 0.109, 185.

14. (i) 0.135, (ii) 0.677; 0.84.

Revision exercise B

1. 6.48 cm, 0.01 cm^2; 22%.

2. (a) 0.40, (b) 0.16.

3. 1.58, 1.26, 0.006.

4. (a) 9, 16.5; (b) 4.5, 107.25.

5. 11, 0.349.

6. (i) 0.55, (ii) 0.18.

7. No; 5.4, 5.4.

8. 63, 65.55, 11.28; Normal.

9. 0.5, 0.5, 0.25, 0.5, 0.5.

10. 0.841, 0.158, 0.9.

11. 7.03, 6.84; $\mu = \sigma^2$ in a Poisson population.

   Sample gives $\mu \simeq \sigma^2$; $64\,e^{-7}\,\dfrac{7^5}{5!}$, 8.17.

12. (i) median $\simeq$ 52.7 kg, (ii) $m \simeq$ 52.51 kg; $s^2 \simeq$ 125.0 kg^2, (iii) 24.9%.

13. (b) (i) $\dfrac{55}{108}$, (ii) $\dfrac{1}{12}$.

14. (i) 0.51,   (ii) 500.98;   5.8.

15. (i) 0.13,   (ii) 0.179, $\begin{array}{c|ccc} & 0 & 1 & 2 \\ \hline & \frac{1}{4} & \frac{1}{2} & \frac{1}{4} \end{array}$ 0.482.

16. 0.90.

17. (i) $P(N = n) = \dfrac{1}{k}$ $n = 1, 2, 3, \ldots, k.$

$E(N) = \dfrac{k + 1}{2},$ (ii) $\dfrac{(k - 1)^{n - 1}}{k^n}$ $n = 1, 2, \ldots,$

$E(N) = k.$

18. (a) 0.135,   (b) 0.271,   (c) 0.029,   (d) 2,   (e) 8.19.

Exercise 12.1

1. (a) 9,   25;   (b) -3,   97;   (c) 6,   4;   (d) 3,   2.34;
   (i) 0.04,   0.62,   0.00,   0.025;
   (ii) 0.025.

2. (a) 0.09,   (b) 0.15.

3. 0.629,   0.357.

4. 3.28.

5. (a) 0.81,   (b) 0.343.

6. (a) 0.067,   (b) 0.17.

7. 0.164,   0.917.

8. (i) 0.023,   (ii) (a) 0.00,   (b) 0.96.

9. (a) 61,   6,   (b) 0.34.

10. (a) 460,   11,   (b) 450 g,   (c) 457 g.

11. (i) (a) 0.309,   (b) 0.067;   (ii) 11.

12. (a) 0.672   (b) 0.520   (c) 0.272.

13. (a) (i) 0.159,   (ii) 0.396;
    (b) (i) 0.739,   (ii) 0.696,   (iii) 0.955.

14. $a_1\mu_1 + a_2\mu_2 + a_3\mu_3,$ $a_1^2\sigma_1^2 + a_2^2\sigma^2 + a_3^2\sigma_3^2;$ 0.0526,   0.1496.

15. 0.13.

16. 0.7, 1.01 ; 0.36, 0.24, 0.16, 0.16, 0.05, 0.02, 0.01, 0.7, 0.505.

17. (i) 0.59,   (ii) 0.20,   (iii) 0.95,   (iv) 0.50.

18. (i) 3,   (ii) 129,   (iii) $N\left(-2, \dfrac{6}{5}\right)$,   (iv) 0.034.

19. 890,   165,   0.73.

20. $\mu = \mu_1 + \mu_2$ (i) 0.091,   (ii) 0.472,   (iii) 0.030.

Exercise 12.2

1. 8.5%.

2. (a) 0.023,   (b) Not a random sample.

3. $N(0, 3.733)$,   0.3.

4. $\mu, \quad \dfrac{\sigma^2}{n_1}; \quad 0, \quad \dfrac{\sigma^2}{n_1} + \dfrac{\sigma^2}{n_2}; \quad 0, \quad 2\sigma^2; \quad 0, \quad 2\dfrac{\sigma^2}{n}.$

5. 0.88.

6. 0.019.

7. 0.2.

8. 21.

9. 0.16.

10. 10,   244;   0.24.

11. (a) 0.68   (b) 1.34, but see section 12.5,
    (c) 1,   (d) 0.68,   0.27.

Exercise 12.3

2. 1.0108, 3.7×10^{-6}, 8.6×10^{-4}.

4. (a) 1.8416, 2.3×10^{-6};
   (b) 1.842, 2.0×10^{-6}.

5. (a) 6.54   (b) 1.528,   (c)   0.553

6. (a) 64,   6.67,   (b) 64,   6.32.

7. 49.75,   0.557

8. 8.7,   0.267.

9. 0.474.

10. $a = b = \dfrac{1}{2}$.

11. (b) $N\left(\mu, \dfrac{1}{n}\right)$, $n \geqslant 97$.

Exercise 12.4

1. (a) -1.87–3.67,   (b) -1.43–3.23.

2. $N\left(\mu, \dfrac{3}{16}\right)$,   3.35–5.05.

3. 9.6,   54.64; Normal with variance 0.5464 and point estimated mean 9.6;   8.15–11.05.

4. -6.1–19.7;   3.6–10.0.

5. 5.9,   1.46;   5.59–6.21.

6. (a) 1600   (b) 385.

7. 341.07–341.33.

8. (21.49, 23.91); sample size at least 349.

9. $2x$, $1.05x - 20x$.

Exercise 13.1

1. $z = -2.93$; a highly significant result.

2. $z = 1.81$; significant with one-tailed test but not with two-tailed.

3. $z = 1.58$; not significant.

4. $z = -3.16$; a highly significant result.

5. $z = 1.26$; not significant.

6. $z = -1.47$; not significant.

7. $z = -1.78$; significant with one-tailed but not two-tailed test.

8. $z = 1.73$; not significant, unless it was thought that bright children had been selected.

9. Not a random selection or perhaps females or children were included.

10. $z = 2$; significant result; yes.

11. $z = -1.75$; not significant.

12. $z = 4.43$; very highly significant; yes.

13. $z = -2.79$; a highly significant result.

14. 0.12, 0.00042; 0.12, 0.000382; not random.

Exercise 13.2

1. $z = \pm 2.38$; significant result.

2. $z = \pm 2.06$; significant result.

3. $z = \pm 0.96$; not significant.

4. $z = \pm 7.72$; very highly significant; yes.

5. $z = \pm 1.13$; not significant.

7. $z = \pm 1.96$; significant result. Had we merely been testing for a difference, this value of z would not have helped. In such cases we would try to obtain more information.

8. $z = -2.41$ a highly significant result; 0.8%.

9. 0.035, 0.235.

10. $z = 4.25$; a very highly significant result.

11. neither significant; $z = 2.3$; significant result; 3.03–3.33.

Exercise 13.3

1. Yes.

2. (a) 0.02,   (b) 0.143,   (c) 0.02,   (d) 0.857.

3. (i) $p^4 + (1 - p)^4$,   (ii) $p^7 + (1 - p)^7 + 7p(1 - p)^6 + 7p^6(1 - p)$;   procedure 2.

4. $P(X = r) = {}^{10}C_r(1 - p)^{10-r} p^r$.

5. Yes;   no change.

6. (b) No (testing at 5% on a one-tailed test);
   (c) Yes, $z = 2$.

Exercise 14.1

1. (a) 0.481,   (b) 0.421.

2. (a) 0.67,   (b) 0.086,   (c) 0.0265.

3. $100\pi, \quad \dfrac{10\,000\,\pi(1 - \pi)}{n}$.

4. (a) 0.49–0.63   (b) 49%–63%.

5. $\dfrac{1}{4}$,   (a) 400,   (b) 256.

6. 0.0041.

7. 66.67,   4.714;   (i) 0.208   (ii) 0.064,   66.67,   3.333;   415.

Exercise 14.2

1. Yes.

2. Yes.

3. 2401.

4. $z = -1.789$;   a significant result.

5. 0.086–0.164 (approx).

6. $z = -2.04$;   Yes.

7. $z = 0.98$;   No.

8. (ii) 0.5 ± 0.024.

9. $z = 2.38$; significant result.

10. 0.510, 0.574; yes.

Revision exercise C

1. 9, 0.613; they should not.

2. (a) 1.5, 1.25; (b) 1.5, 0.625.

3. 0.344, 0.0625.

5. 26.75, 42.02; not significant at 5% level.

6. 2.51, 1.34, 0.946; Use $\text{Var}(X + Y) = \text{Var}(X) + \text{Var}(Y)$.

7. 3, 0.25, 0.1936; 12, 0.6, 0.2;
   (a) 0.125,   (b) 0.3125.

8. $\bar{x} \simeq 49.9$, $s \simeq 13.7$.

9. (a) 0.62   (b) 0.38.

10. $P(X = 6) = \dfrac{3}{36}$, $P(X = 7) = \dfrac{12}{36}$, $P(X = 8) = \dfrac{12}{36}$, $P(X = 9) = \dfrac{8}{36}$;

   $\sigma \simeq 0.97$; $\dfrac{829}{1296}$.

11. $-6.324 - 5.436$; could be zero.

12. 0.322, 50p, 50, 21.

14. θ, $\sqrt{\dfrac{\theta(1 - \theta)}{n}}$, 0.83–0.94.

15. $v = \dfrac{1.645}{\sqrt{n}}$, $n \geqslant 215$.

16. 0.159, 225; 1200, 1225, 1282; 0.127.

17. (a) 0.878, (b) 9.69, 0.2339 (c) 0.180.

18. 0.371, £60.37.

19. (a) $\dfrac{21}{40}$ (b) $\dfrac{19}{30}$ (c) 0.400 ± 0.048.

20. $4\frac{1}{12}$, $\dfrac{6}{17}$.

21.

0	1	2	3
$\frac{1}{12}$	$\frac{5}{12}$	$\frac{5}{12}$	$\frac{1}{12}$

1.5, 0.583.

Exercise 16.1
1. 3.85–156.4.

2. $\chi^2(5) = 5.2$; no evidence against the hypothesis.

3. $\chi^2(9) = 11$; no evidence against the hypothesis.

4. $\chi^2(2) = 22.26$; no.

5. $\chi^2(1) = 1.74$; no (remember to group).

6. $\chi^2(3) = 4.71$; no.

7. (a) $\chi^2(2) = 12.7$; no; (b) $\chi^2(2) = 1.2$; no evidence against the hypothesis.

9. (a) $\chi^2(16) = 11.8$; no evidence against the hypothesis.

   (b) $\chi^2(5) = 4.1$; no evidence against the hypothesis.

11. $\chi^2(2) = 3.23$; yes.

12. (a) $\chi^2(4) = 16.9$; not likely;

    (b) $\chi^2(4) = 0.8$; no evidence against the hypothesis.

Exercise 16.2
1. $\chi^2(2) = 7.4$; significant result; not independent.

2. $\chi^2(9) = 20.7$; yes.

3. $\chi^2(1) = 1.13$; no (remember Yates' correction).

4. $\chi^2(1) = 0.027$; no evidence against the hypothesis.

5. $\chi^2(2) = 13.7$; a highly significant result.

6. $\chi^2(1) = 0.088$; no significant difference.

Exercise 17.1

2. (a) 1,   (b) 0,   (c) 0.

7. 0.9063.

8. 0.906.

9. (a) 0.081,   (b) -0.386,   (c) -0.07.

Exercise 17.2

1. $r_s = 0.83$;   yes.

2. $r_s = 0.12$.

3. (a) -0.374,   (b) -0.262,   (c) no.

4. $r_s(PQ) = 0.75$; $r_s(QR) = 0.567$;
   $r_s(PR) = 0.317$; P and Q.

5. 0.488,   0.8303.

Exercise 18.1

1. $y = 1.486x - 1.057$.

2. $y = 11.03 - 0.817x$, $x = 12.755 - 1.096x$,   6.18.

3. $y = 10.59 - 1.167x$.

4. (a) 0.63,   (b) $x = 0.35y + 1.212$.

5. (a) $y = 21.6 - 3.6x$,   (b) $y = 7.2 - 1.2w$.

7. (a) -180.4,   (b) -0.926,   (c) $y = 58.73 - 0.752$; covariance depends on units,
   r does not.

8. $y = 0.103x + 3.99$,   9.55.

9. $y = 0.652x - 2.59$,   36.5.

10. (i) $y = 0.785x + 20.526$.
    (ii) $x = 1.117y - 11.889$;   0.936.

11. (a) $y = 0.486x - 2.397$.
    (b) 0.997,   (c) No.

12. $y = 0.53x - 25.804$,
    $x = 1.008y + 94.384$,
    0.73; $53.7m$.

13. (b) $y = 2.75x + 48.35$,   (c) 0.79.

14. Many reasons; e.g., points appear to lie on a line, or there may be a theory to
    support linearity. Choose m so that $Y = mX$ is 'best fit'. Least squares gives $m = \Sigma XY/\Sigma X^2$; $m = 3.34$.

15. $y = 0.9x - 3.353$;   (a) 33,   (b) 15.

16. $x + 8y = $ const.; $y = -0.6996x - 0.0025$.
    (a) $\gamma \simeq 1.43$,   (b) $V \simeq 1.22\, m^3$.
    The range of p should include $0.75\, \mathrm{Nm^{-2}}$ to prevent extrapolation.

17. $z = 0.02715x + 0.9696$.

18. $H = 1.158A + 107.31$.

19. $y = 0.233x + 11.473$.

20. $y = 1.284x$;   5.39 km.

Exercise 19.1

1. (a) $\dfrac{1}{4}$,   (b) $\dfrac{\pi}{2}$,   (c) $\dfrac{\pi^2}{4}$.

2. $\dfrac{1}{\pi}$, $\dfrac{1}{16}$.

3. $\dfrac{1}{20}$,   20,   (a) 0.049,   (b) 0.306.

4. (a) 1.386,   (b) 0.078,   (c) 0.443,   (d) $\dfrac{4}{3}$,   (e) 0.

6. (i) $\displaystyle \int_{-\infty}^{a} \frac{1}{\sigma\sqrt{2\pi}}\, e^{[-\frac{1}{2}(x-\mu/\sigma)^2]}\, dx$

7. (a) (i) $\dfrac{1}{2}(a + b)$, $\dfrac{1}{2}(a^2 + ab + b^2)$

   $\dfrac{1}{2}(a + b)$, $\dfrac{1}{12}(b - a)^2$

   (ii) $\dfrac{1}{2}(N + 1)$, $\dfrac{1}{6}(N + 1)(2N + 1)$, $\dfrac{1}{2}(N + 1)$, $\dfrac{1}{12}(N^2 - 1)$.

8. 12, 0.4, 0.04, $\frac{1}{3}$.

9. (a) $\frac{1}{2}$, (b) $\frac{1}{2}$.

10. $\frac{1}{\sqrt{128\,\pi}}\,e^{[-(x-50)^2/128]}$ 0.106;   prop. = 0.734,

   $A = 1.36$;   0.144.

11. $e^{-4} = 0.018$, $(1 - e^{-4})^{n-1}\,e^{-4}$.

12. μ, σ^2; $N\left(\mu, \dfrac{4}{n}\right)$; $n \geqslant 62$.

13. $\dfrac{2}{n(n+1)}$, $\dfrac{1}{3}(2n+1)$,   8.5.

14. 2, $\dfrac{3}{4}$;   0.2.

Exercise 19.2

2. $\frac{1}{3}V^{-2/3}$, $1 \leqslant v \leqslant 8$;   3.75,   4.08.

3. (a) $\dfrac{1}{6\pi}$ for $2\pi \leqslant y \leqslant 8\pi$,   (b) $\dfrac{1}{6\sqrt{\pi}}z^{-1/2}$, $\pi \leqslant z \leqslant 16\,\pi$.

4 $4a/\pi$, $2a^2 - \dfrac{16a^2}{\pi^2}$.

5. $\dfrac{2}{3}$, $r\sqrt{2}$, $\dfrac{2}{\pi}\sin^{-1}\dfrac{k}{2r}$.

6. 8.67,   21.42.

7. 0.6, $4x^3 - 3x^4$.

8. 3π, $3\pi^2$,   0.63.

9. $f_H(h) = \dfrac{\pi h^2}{3}$, $0 \leqslant h \leqslant 1.42$.

   (a) 0.349,   (b) 1.064,   (c) 0.077,   (d) 0.420.

10. λ,   100,   10 000,   0.607,   0.368.

12. $\dfrac{a}{2}, \quad \dfrac{a\sqrt{3}}{6}, \quad 1 - \dfrac{(2a - y)^2}{2a^2},$

$$f_Y(y) = \begin{cases} \dfrac{y}{a^2}, & 0 \leqslant y \leqslant a \\[2mm] \dfrac{2}{a} - \dfrac{y}{a^2}, & a < y \leqslant 2a \\[2mm] 0, & \text{otherwise.} \end{cases}$$

13. (c)

$$F(x) = \begin{cases} -\dfrac{1}{n}x^3, & x \leqslant -1 \\[2mm] \dfrac{1}{2} + \dfrac{1}{2}x - \dfrac{1}{12}x^3, & -1 \leqslant x \leqslant 1 \\[2mm] 1 - \dfrac{1}{12}x^3, & x \geqslant 1 \end{cases}$$

(d) 0, $\dfrac{11}{15}$.

14. (a) $\dfrac{1}{2}, \quad -\dfrac{1}{4}$ (b) $\dfrac{7}{16}$.

(c)

$$f(x) = \begin{cases} \dfrac{1}{2x}, & 0 \leqslant x \leqslant 1 \\[2mm] \dfrac{1}{2}, & 1 < x \leqslant 2 \\[2mm] -\dfrac{1}{2}x + \dfrac{3}{2}, & 2 < x \leqslant 3 \\[2mm] 0, & \text{otherwise} \end{cases}$$

$\dfrac{3}{2}, \quad \dfrac{5}{12}$.

Revision exercise D

1. (a) 210, (i) $\dfrac{1}{14}$, (ii) $\dfrac{3}{7}$, (iii) $\dfrac{1}{7}$, (iv) $\dfrac{4}{15}$;

(b) $P(A) = \dfrac{9}{25}, \quad P(B) = \dfrac{12}{25}, \quad P(C) = \dfrac{4}{25}$.

2. $\dfrac{1}{6}, \quad \dfrac{1}{6}, \quad \dfrac{1}{36}, \quad \dfrac{1}{36}, \quad \dfrac{1}{36}, \quad \dfrac{1}{6}$.

3. $\dfrac{3}{16}$, $\dfrac{21}{128}$, $\dfrac{70}{256}$, $\dfrac{115}{512}$.

4. (i) $\dfrac{8}{9}$, (ii) $\dfrac{1}{9}$, (iii) $\dfrac{1}{9}$, (iv) $\dfrac{1}{63}$, (v) $0.206 = \dfrac{13}{65}$,

   (vi) $\dfrac{50}{63} = 0.794$, (vii) $\dfrac{2}{63} = 0.032$.

5. (a) 0.271, (b) 0.252, (c) 0.122, (d) 0.279.

6. $P(A \cup B \cup C) = P(A) + P(B) + P(C) - P(A \cap C) - P(A \cap B)$
$$- P(B \cap C) + P(A \cap B \cap C).$$

7. (a) 0.88, (b) 0.792, (c) 0.512; 0.894.

8. $\dfrac{7}{16}$.

9. $^{2n}C_n(\tfrac{1}{2})^{2n}$.

10. 0.158.

11. (a) (i) $\dfrac{1}{2}$, (ii) $\dfrac{11}{36}$, (iii) $\dfrac{1}{6}$, (iv) $\dfrac{23}{36}$, (v) $\dfrac{6}{11}$;

   (b) (i) 0.328, (ii) 0.081, (iii) $n \geqslant 22$.

12. (a) (i) $\dfrac{2}{17}$, (ii) $\dfrac{11}{850}$, (iii) $\dfrac{1}{5525}$; $\dfrac{1}{4606}$;

   (b) $\dfrac{5}{9}$.

13. Use $B(6, 0.6)$, 3.6 (i) 0.456, (ii) 0.233, 2.

14.

2	3	4
$p_1 p_2 + (1 - p_1)(1 - p_1)$	$p_1(1 - p_2) + (1 - p_1)p_1 p_2$	$p_1(1 - p_1)(1 - p_2)$

$$p = \dfrac{4 - \sqrt{7}}{3}.$$

15. (a) $\dfrac{1}{32} = 0.0312$, (b) $\dfrac{13}{20} = 0.65$, (c) $\dfrac{17}{125} = 0.136$,

   (d) $\dfrac{24}{125} = 0.192$, (e) $\dfrac{1}{64} = 0.016$.

16. (a) $\beta + \dfrac{1}{5}(\alpha - \beta)(\alpha + 4\beta)$,   (c) X and Y are independent events.

17. $q = \dfrac{1}{25}$,   0.022;   £760.

18. $P(X = r) = \left(\dfrac{2}{3}\right)^{r-1}\dfrac{1}{3}$, $r = 1, 2, 3, \ldots$.

$P(X = r) = \dfrac{1}{3}$, $r = 1, 2, 3$,   (a) $\dfrac{8}{27}$,   (b) $\dfrac{38}{81}$.

19. (a) (i) 0.135,   (ii) 0.323,   (b) 0.250.

20. (d) $\dfrac{24}{625}$,   (b) $\dfrac{24}{625}$ (c) 2.

Index